対訳
ISO 19011:2018
(JIS Q 19011:2019)

ポケット版

マネジメントシステム監査のための指針

日本規格協会　編

2019年7月1日のJIS法改正により名称が変わりました．本書に収録しているJISについても，まえがきを除き，規格中の「日本工業規格」を「日本産業規格」に読み替えてください．

＊著作権について

本書は，ISO中央事務局と当会との翻訳出版契約に基づいて刊行したものです．

本書に収録したISO及びJISは，著作権により保護されています．本書の一部又は全部について，当会及びISOの許可なく複写・複製することを禁じます．ISOの著作権は，下に示すとおりです．

本書の著作権に関するお問合せは，日本規格協会グループ販売サービスチーム（Tel. 03-4231-8550）にて承ります．

© ISO 2018

All rights reserved. Unless otherwise specified, or required in the context of its implementation, no part of this publication may be reproduced or utilized otherwise in any form or by any means, electronic or mechanical, including photocopying, or posting on the internet or an intranet, without prior written permission. Permission can be requested from either ISO at the address below or ISO's member body in the country of the requester.

    ISO copyright office
    CP 401・Ch. de Blandonnet 8
    CH-1214 Vernier, Geneva
    Phone: +41 22 749 01 11
    Fax: +41 22 749 09 47
    Email: copyright@iso.org
    Website: www.iso.org
Published in Switzerland

本書について

　本書は，国際標準化機構（ISO）が 2018 年 7 月に第 3 版として発行した国際規格 ISO 19011:2018（Guidelines for auditing management systems），及びその翻訳規格として，日本工業標準調査会（JISC）の審議を経て 2019 年 5 月 20 日に経済産業大臣が改正した日本工業規格（現 日本産業規格）JIS Q 19011:2019（マネジメントシステム監査のための指針）を，英和対訳で収録したものです．

　収録に際して JIS の解説は省略しています．JIS の解説を参照したい場合は，JIS 規格票をご利用ください．

　規格をより深く理解したい方には，『ISO 19011:2018（JIS Q 19011:2019）マネジメントシステム監査　解説と活用方法』（日本規格協会，2019）をはじめとする関係書籍を併読されることをお勧めします．

2019 年 8 月

日本規格協会

Contents

ISO 19011:2018
Guidelines for auditing management systems

Foreword ·· 8
Introduction ·· 16
1 **Scope** ··· 28
2 **Normative references** ································· 30
3 **Terms and definitions** ································ 30
4 **Principles of auditing** ································· 58
5 **Managing an audit programme** ················· 66
5.1 General ··· 66
5.2 Establishing audit programme objectives ··· 76
5.3 Determining and evaluating audit programme risks and opportunities ········· 80
5.4 Establishing the audit programme ········· 84
5.5 Implementing audit programme ············· 98
5.6 Monitoring audit programme ················· 124
5.7 Reviewing and improving audit programme ··· 128
6 **Conducting an audit** ···································· 130

目　次

JIS Q 19011:2019
マネジメントシステム監査のための指針

まえがき …………………………………… 9
序文 ………………………………………… 17
1　適用範囲 ………………………………… 29
2　引用規格 ………………………………… 31
3　用語及び定義 …………………………… 31
4　監査の原則 ……………………………… 59
5　監査プログラムのマネジメント ……… 67
5.1　一般 …………………………………… 67
5.2　監査プログラムの目的の確立 ……… 77

5.3　監査プログラムのリスク及び機会の
　　 決定及び評価 ………………………… 81
5.4　監査プログラムの確立 ……………… 85
5.5　監査プログラムの実施 ……………… 99
5.6　監査プログラムの監視 ……………… 125
5.7　監査プログラムのレビュー及び改善 … 129

6　監査の実施 ……………………………… 131

6.1	General	130
6.2	Initiating audit	132
6.3	Preparing audit activities	138
6.4	Conducting audit activities	152
6.5	Preparing and distributing audit report	188
6.6	Completing audit	194
6.7	Conducting audit follow-up	196
7	**Competence and evaluation of auditors**	198
7.1	General	198
7.2	Determining auditor competence	202
7.3	Establishing auditor evaluation criteria	226
7.4	Selecting appropriate auditor evaluation method	226
7.5	Conducting auditor evaluation	230
7.6	Maintaining and improving auditor competence	232

Annex A (informative) **Additional guidance for auditors planning and conducting audits** 236

Bibliography 300

6.1	一般	131
6.2	監査の開始	133
6.3	監査活動の準備	139
6.4	監査活動の実施	153
6.5	監査報告書の作成及び配付	189
6.6	監査の完了	195
6.7	監査のフォローアップの実施	197

7 監査員の力量及び評価 ································ 199

7.1	一般	199
7.2	監査員の力量の決定	203
7.3	監査員の評価基準の確立	227
7.4	監査員の適切な評価方法の選択	227
7.5	監査員の評価の実施	231
7.6	監査員の力量の維持及び向上	233

附属書 A（参考） 監査を計画及び実施する監査員に対する追加の手引 ···················· 237

参考文献 ···································· 301

Foreword

ISO (the International Organization for Standardization) is a worldwide federation of national standards bodies (ISO member bodies). The work of preparing International Standards is normally carried out through ISO technical committees. Each member body interested in a subject for which a technical committee has been established has the right to be represented on that committee. International organizations, governmental and non-governmental, in liaison with ISO, also take part in the work. ISO collaborates closely with the International Electrotechnical Commission (IEC) on all matters of electrotechnical standardization.

The procedures used to develop this document and those intended for its further maintenance are described in the ISO/IEC Directives, Part 1. In particular the different approval criteria needed for the different types of ISO documents should be noted. This document was drafted in accordance

まえがき

(ISO の Foreword と JIS のまえがきは,それぞれの原文において内容が異なっているため,対訳となっていないことにご注意ください.)

この規格は,工業標準化法第 14 条によって準用する第 12 条第 1 項の規定に基づき,一般財団法人日本規格協会(JSA)から,工業標準原案を具して日本工業規格を改正すべきとの申出があり,日本工業標準調査会の審議を経て,経済産業大臣が改正した日本工業規格である.

これによって,**JIS Q 19011**:2012 は改正され,この規格に置き換えられた.

この規格は,著作権法で保護対象となっている著作物である.

この規格の一部が,特許権,出願公開後の特許出願又は実用新案権に抵触する可能性があることに注意を喚起する.経済産業大臣及び日本工業標準調査会は,このような特許権,出願公開後の特許出願及び実用新案権に関わる確認について,責任はもたない.

with the editorial rules of the ISO/IEC Directives, Part 2 (see **www.iso.org/directives**).

Attention is drawn to the possibility that some of the elements of this document may be the subject of patent rights. ISO shall not be held responsible for identifying any or all such patent rights. Details of any patent rights identified during the development of the document will be in the Introduction and/or on the ISO list of patent declarations received (see **www.iso.org/patents**).

Any trade name used in this document is information given for the convenience of users and does not constitute an endorsement.

For an explanation on the voluntary nature of standards, the meaning of ISO specific terms and expressions related to conformity assessment, as well as information about ISO's adherence to the World Trade Organization (WTO) principles in the Technical Barriers to Trade (TBT) see the following URL: **www.iso.org/iso/foreword.html**.

This document was prepared by Project Committee ISO/PC 302, *Guidelines for auditing management systems*.

This third edition cancels and replaces the second edition (ISO 19011:2011), which has been technically revised.

The main differences compared to the second edition are as follows:
— addition of the risk-based approach to the principles of auditing;
— expansion of the guidance on managing an audit programme, including audit programme risk;
— expansion of the guidance on conducting an audit, particularly the section on audit planning;
— expansion of the generic competence requirements for auditors;
— adjustment of terminology to reflect the process and not the object ("thing");
— removal of the annex containing competence requirements for auditing specific manage-

ment system disciplines (due to the large number of individual management system standards, it would not be practical to include competence requirements for all disciplines);
— expansion of **Annex A** to provide guidance on auditing (new) concepts such as organization context, leadership and commitment, virtual audits, compliance and supply chain.

Introduction

Since the second edition of this document was published in 2011, a number of new management system standards have been published, many of which have a common structure, identical core requirements and common terms and core definitions. As a result, there is a need to consider a broader approach to management system auditing, as well as providing guidance that is more generic. Audit results can provide input to the analysis aspect of business planning, and can contribute to the identification of improvement needs and activities.

序文

(ISO の Introduction と JIS の序文は，それぞれの原文において内容が異なっているため，対訳となっていないことにご注意ください.)

この規格は，2018 年に第 3 版として発行された **ISO 19011** を基に，技術的内容及び構成を変更することなく作成した日本工業規格である.

なお，この規格で点線の下線を施してある参考事項は，対応国際規格にはない事項である.

この規格の 2012 年版に対する主な変更点は，次のとおりである.

— 監査の原則への，リスクに基づくアプローチの追加
— 監査プログラムのマネジメントに関する手引の拡張. この拡張には監査プログラムのリスクを含む.
— 監査の実施に関する手引の拡張，特に，監査計画の策定の部分
— 監査員に関する共通的な力量要求事項の拡張
— 対象［"もの (thing)"）］でなく，プロセスを反映する用語の調整
— 特定のマネジメントシステム分野の監査に関する力量要求事項を扱う附属書の削除（個々のマネジメントシステム規格の数が多く，全ての分

An audit can be conducted against a range of audit criteria, separately or in combination, including but not limited to:
— requirements defined in one or more management system standards;
— policies and requirements specified by rele-

野に関する力量要求事項を含めるのは現実的でない.)
― 組織構造, リーダーシップ及びコミットメント, 仮想監査, 順守, サプライチェーンなどの (新しい) 概念を監査することに関する手引を提供するための**附属書 A** の拡張

JIS Q 19011:2012 を発効して以降, 多くの新しいマネジメントシステム規格が発効されてきており, その多くが共通の構造, 共通の中核となる要求事項, 並びに共通の用語及び中核となる定義をもっている. 結果として, より共通的な手引を与えることに加え, マネジメントシステム監査へのより幅広いアプローチを考慮する必要がある. 監査結果は, 事業計画策定の分析の側面に対してインプットを提供し, 改善の必要性及び活動の特定に寄与することができる.

監査は, 様々な監査基準の, 個別又は組合せに対して行うことができる. この監査基準には次の事項を含むが, これらに限らない.
― 一つ又は複数のマネジメントシステム規格で定める要求事項
― 関連する利害関係者が規定する方針及び要求事

vant interested parties;
- statutory and regulatory requirements;
- one or more management system processes defined by the organization or other parties;
- management system plan(s) relating to the provision of specific outputs of a management system (e.g. quality plan, project plan).

This document provides guidance for all sizes and types of organizations and audits of varying scopes and scales, including those conducted by large audit teams, typically of larger organizations, and those by single auditors, whether in large or small organizations. This guidance should be adapted as appropriate to the scope, complexity and scale of the audit programme.

This document concentrates on internal audits (first party) and audits conducted by organizations on their external providers and other external interested parties (second party). This document can also be useful for external audits conducted for purposes other than third party management system certification. ISO/IEC 17021-1 provides re-

項
— 法令・規制要求事項
— 組織又は他の関係者が定めた一つ又は複数のマネジメントシステムプロセス
— マネジメントシステムの特定のアウトプットの提供に関係するマネジメントシステムの計画(例えば,品質計画,プロジェクト計画)

この規格は,全ての規模及びタイプの組織,並びに様々な範囲及び規模の監査に対して,手引を提供する.これには,一般的に更に大規模な組織で大規模監査チームが行う監査,及び組織規模の大小に関わりなく単独の監査員が行う監査が含まれる.この手引は,監査プログラムの範囲,複雑性及び規模に適切に対応させることが望ましい.

この規格は,内部監査(第一者),並びに組織が組織の外部提供者及びその他の外部利害関係者(第二者)に対して行う監査に焦点を合わせている.この規格はまた,第三者マネジメントシステム認証以外の目的で行う外部監査においても有用となり得る.**JIS Q 17021-1** は,認証を目的としたマネジメントシステムの監査における要求事項を提供す

quirements for auditing management systems for third party certification; this document can provide useful additional guidance (see **Table 1**).

Table 1 — **Different types of audits**

1st party audit	2nd party audit	3rd party audit
Internal audit	External provider audit	Certification and/or accreditation audit
	Other external interested party audit	Statutory, regulatory and similar audit

To simplify the readability of this document, the singular form of "management system" is preferred, but the reader can adapt the implementation of the guidance to their own situation. This also applies to the use of "individual" and "individuals", "auditor" and "auditors".

This document is intended to apply to a broad range of potential users, including auditors, organizations implementing management systems and organizations needing to conduct management system audits for contractual or regulatory reasons. Users of this document can, however, apply this guidance in developing their own audit-related requirements.

る．この規格は，有用な追加的な手引を提供し得る（**表1**参照）．

表1－監査のタイプ

第一者監査	第二者監査	第三者監査
内部監査	外部提供者監査	認証審査及び／又は認定監査
	他の外部利害関係者による監査	法令，規制及び類似の監査

　この規格は，幅広い潜在的利用者に適用することを意図している．この潜在的利用者には，監査員，マネジメントシステムを実施する組織，及び契約上の又は規制上の理由によってマネジメントシステム監査の実施が必要な組織を含む．一方で，この規格の利用者は，利用者自身の監査に関連する要求事項を作成するときにこの手引を適用することができる．

The guidance in this document can also be used for the purpose of self-declaration and can be useful to organizations involved in auditor training or personnel certification.

The guidance in this document is intended to be flexible. As indicated at various points in the text, the use of this guidance can differ depending on the size and level of maturity of an organization's management system. The nature and complexity of the organization to be audited, as well as the objectives and scope of the audits to be conducted, should also be considered.

This document adopts the combined audit approach when two or more management systems of different disciplines are audited together. Where these systems are integrated into a single management system, the principles and processes of auditing are the same as for a combined audit (sometimes known as an integrated audit).

This document provides guidance on the management of an audit programme, on the planning and

この規格に示す手引はまた，自己宣言のために利用することもでき，さらに，監査員の研修機関又は要員認証機関にとっても有用となり得る．

　この規格に示す手引は，柔軟性のあることを意図している．規格本文の様々な箇所で示すように，この手引の利用の仕方は，監査の対象となる組織のマネジメントシステムの規模及び成熟度に応じて変わり得る．また，監査の対象となる組織の性質及び複雑さ，並びに実施する監査の目的及び範囲も考慮することが望ましい．

　この規格は，異なった分野の複数のマネジメントシステムを一緒に監査する場合，複合監査のアプローチを取り入れている．これらのシステムを一つのマネジメントシステムに統合する場合，監査の原則及びプロセスは，複合監査（統合監査といわれることもある）の場合と同じである．

　この規格は，監査プログラムのマネジメント，マネジメントシステム監査の計画及び実施，並びに監

conducting of management system audits, as well as on the competence and evaluation of an auditor and an audit team.

査員及び監査チームの力量及び評価に関する手引を提供している．

1 Scope

This document provides guidance on auditing management systems, including the principles of auditing, managing an audit programme and conducting management system audits, as well as guidance on the evaluation of competence of individuals involved in the audit process. These activities include the individual(s) managing the audit programme, auditors and audit teams.

It is applicable to all organizations that need to plan and conduct internal or external audits of management systems or manage an audit programme.

The application of this document to other types of audits is possible, provided that special consideration is given to the specific competence needed.

1 適用範囲

この規格は，マネジメントシステム監査のための手引を提供する．これには，監査の原則，監査プログラムのマネジメント，マネジメントシステム監査の実施，並びに監査プロセスに関わる人の力量の評価を含む．これらの活動には，監査プログラムをマネジメントする人，監査員及び監査チームを含む．

この規格は，マネジメントシステムの内部監査若しくは外部監査を計画し，行う必要のある，又は監査プログラムのマネジメントを行う必要のある全ての組織に適用できる．

この規格を他のタイプの監査に適用することも可能ではあるが，その場合は，必要とされる固有の力量について特別な考慮が必要となる．

注記　この規格の対応国際規格及びその対応の程度を表す記号を，次に示す．

ISO 19011:2018, Guidelines for auditing management systems（IDT）

なお，対応の程度を表す記号"IDT"は，**ISO/IEC Guide 21-1** に基づき，

2 Normative references

There are no normative references in this document.

3 Terms and definitions

For the purposes of this document, the following terms and definitions apply.

ISO and IEC maintain terminological databases for use in standardization at the following addresses:

— ISO Online browsing platform: available at **https://www.iso.org/obp**
— IEC Electropedia: available at **http://www.electropedia.org/**

3.1
audit
systematic, independent and documented process for obtaining *objective evidence* (**3.8**) and evaluating it objectively to determine the extent to which the *audit criteria* (**3.7**) are fulfilled

"一致している"ことを示す.

2 引用規格
この規格には,引用規格はない.

3 用語及び定義
この規格で用いる主な用語の定義は,次による.

ISO 及び IEC は,標準化に使用するための用語データベースを次のアドレスに維持している.

— **ISO** Online browsing platform:https://www.iso.org/obp
— **IEC** Electropedia:http://www.electropedia.org/

3.1
監査(audit)
監査基準(**3.7**)が満たされている程度を判定するために,**客観的証拠**(**3.8**)を収集し,それを客観的に評価するための,体系的で,独立し,文書化したプロセス.

Note 1 to entry: Internal audits, sometimes called first party audits, are conducted by, or on behalf of, the organization itself.

Note 2 to entry: External audits include those generally called second and third party audits. Second party audits are conducted by parties having an interest in the organization, such as customers, or by other individuals on their behalf. Third party audits are conducted by independent auditing organizations, such as those providing certification/registration of conformity or governmental agencies.

[SOURCE: ISO 9000:2015, 3.13.1, modified — Notes to entry have been modified]

3.2
combined audit
audit (**3.1**) carried out together at a single *auditee* (**3.13**) on two or more *management systems* (**3.18**)

Note 1 to entry: When two or more discipline-specific management systems are integrated into

注記 1 内部監査は,第一者監査と呼ばれることもあり,その組織自体又は代理人によって行われる.

注記 2 外部監査には,一般的に第二者監査及び第三者監査と呼ばれるものが含まれる.第二者監査は,顧客など,その組織に利害をもつ者又はその代理人によって行われる.第三者監査は,適合に関する認証・登録を提供する機関又は政府機関のような,独立した監査組織によって行われる.

(出典:**JIS Q 9000**:2015 の **3.13.1** を変更.注記を変更した.)

3.2
複合監査(combined audit)
　一つの**被監査者**(**3.13**)において,複数の**マネジメントシステム**(**3.18**)を同時に**監査**(**3.1**)すること.

　　注記　複数の分野固有のマネジメントシステムを単一のマネジメントシステムに統合す

a single management system this is known as an integrated management system.

[SOURCE: ISO 9000:2015, 3.13.2, modified]

3.3
joint audit
audit (**3.1**) carried out at a single *auditee* (**3.13**) by two or more auditing organizations

[SOURCE: ISO 9000:2015, 3.13.3]

3.4
audit programme
arrangements for a set of one or more *audits* (**3.1**) planned for a specific time frame and directed towards a specific purpose

[SOURCE: ISO 9000:2015, 3.13.4, modified — wording has been added to the definition]

3.5
audit scope
extent and boundaries of an *audit* (**3.1**)

る場合，これは統合マネジメントシステムと呼ばれる．

（出典：**JIS Q 9000**:2015 の **3.13.2** を変更．）

3.3
合同監査（joint audit）

複数の**監査**（**3.1**）する組織が一つの**被監査者**（**3.13**）を監査すること．

（出典：**JIS Q 9000**:2015 の **3.13.3**）

3.4
監査プログラム（audit programme）

特定の目的に向けた，決められた期間内で実行するように計画された一連の**監査**（**3.1**）に関する取決め．

（出典：**JIS Q 9000**:2015 の **3.13.4** を変更．定義に語句を追加した．）

3.5
監査範囲（audit scope）

監査（**3.1**）の及ぶ領域及び境界．

Note 1 to entry: The audit scope generally includes a description of the physical and virtual-locations, functions, organizational units, activities and processes, as well as the time period covered.

Note 2 to entry: A virtual location is where an organization performs work or provides a service using an on-line environment allowing individuals irrespective of physical locations to execute processes.

[SOURCE: ISO 9000:2015, 3.13.5, modified — Note 1 to entry has been modified, Note 2 to entry has been added]

3.6
audit plan
description of the activities and arrangements for an *audit* (**3.1**)

[SOURCE: ISO 9000:2015, 3.13.6]

3.7
audit criteria

> **注記 1** 監査範囲には,一般に,物理的及び仮想的な場所,機能,組織単位,活動,プロセス,並びに監査の対象となる期間を示すものを含む.
>
> **注記 2** 仮想的な場所とは,オンライン環境を用いて,組織が作業を実施する,又はサービスを提供する場所のことであり,その環境では,人が物理的な場所にかかわらずプロセスを実行することを可能にする.

(出典:**JIS Q 9000**:2015 の **3.13.5** を変更.**注記 1** を変更し,**注記 2** を追加した.)

3.6
監査計画(audit plan)

監査(**3.1**)のための活動及び手配事項を示すもの.

(出典:**JIS Q 9000**:2015 の **3.13.6**)

3.7
監査基準(audit criteria)

set of *requirements* (**3.23**) used as a reference against which *objective evidence* (**3.8**) is compared

Note 1 to entry: If the audit criteria are legal (including statutory or regulatory) requirements, the words "compliance" or "non-compliance" are often used in an *audit finding* (**3.10**).

Note 2 to entry: Requirements may include policies, procedures, work instructions, legal requirements, contractual obligations, etc.

[SOURCE: ISO 9000:2015, 3.13.7, modified — the definition has been changed and Notes to entry 1 and 2 have been added]

3.8
objective evidence
data supporting the existence or verity of something

Note 1 to entry: Objective evidence can be obtained through observation, measurement, test or by other means.

Note 2 to entry: Objective evidence for the purpose

客観的証拠（3.8）と比較する基準として用いる一連の**要求事項**（3.23）．

> 注記1　監査基準が法的（法令・規制を含む．）要求事項である場合，**監査所見**（3.10）において"順守"又は"不順守"の用語がしばしば用いられる．
>
> 注記2　要求事項には，方針，手順，作業指示，法的要求事項，契約上の義務などを含んでもよい．

（出典：**JIS Q 9000**:2015の**3.13.7**を変更．定義を変更し，**注記**1及び**注記**2を追加した．）

3.8
客観的証拠（objective evidence）

あるものの存在又は真実を裏付けるデータ．

> 注記1　客観的証拠は，観察，測定，試験又はその他の手段によって得ることができる．
>
> 注記2　**監査**（3.1）のための客観的証拠は，

of the *audit* (**3.1**) generally consists of records, statements of fact, or other information which are relevant to the *audit criteria* (**3.7**) and verifiable.

[SOURCE: ISO 9000:2015, 3.8.3]

3.9
audit evidence
records, statements of fact or other information, which are relevant to the *audit criteria* (**3.7**) and verifiable

[SOURCE: ISO 9000:2015, 3.13.8]

3.10
audit findings
results of the evaluation of the collected *audit evidence* (**3.9**) against *audit criteria* (**3.7**)

Note 1 to entry: Audit findings indicate *conformity* (**3.20**) or *nonconformity* (**3.21**).
Note 2 to entry: Audit findings can lead to the identification of risks, opportunities for improvement or recording good practices.

一般に，**監査基準**（**3.7**）に関連し，かつ，検証できる，記録，事実の記述又はその他の情報から成る．

（出典：**JIS Q 9000**:2015 の **3.8.3**）

3.9
監査証拠（audit evidence）
　監査基準（**3.7**）に関連し，かつ，検証できる，記録，事実の記述又はその他の情報．

（出典：**JIS Q 9000**:2015 の **3.13.8**）

3.10
監査所見（audit findings）
　収集された**監査証拠**（**3.9**）を，**監査基準**（**3.7**）に対して評価した結果．

　　注記 1　監査所見は，**適合**（**3.20**）又は**不適合**（**3.21**）を示す．
　　注記 2　監査所見は，リスク若しくは改善の機会の特定，又は優れた実践事例の記録を導き得る．

Note 3 to entry: In English if the audit criteria are selected from statutory requirements or regulatory requirements, the audit finding is termed compliance or non-compliance.

[SOURCE: ISO 9000:2015, 3.13.9, modified — Notes to entry 2 and 3 have been modified]

3.11
audit conclusion

outcome of an *audit* (**3.1**), after consideration of the audit objectives and all *audit findings* (**3.10**)

[SOURCE: ISO 9000:2015, 3.13.10]

3.12
audit client

organization or person requesting an *audit* (**3.1**)

Note 1 to entry: In the case of internal audit, the audit client can also be the *auditee* (**3.13**) or the individual(s) managing the audit programme. Requests for external audit can come from sources such as regulators, contracting parties or potential

注記 3　監査基準が法令要求事項又は規制要求事項から選択される場合，監査所見は"順守"又は"不順守"と呼ばれる．

(出典：**JIS Q 9000**:2015 の **3.13.9** を変更．**注記 2** 及び**注記 3** を変更した．)

3.11
監査結論（audit conclusion）

監査（3.1）目的及び全ての**監査所見**（**3.10**）を考慮した上での，監査の結論．

(出典：**JIS Q 9000**:2015 の **3.13.10**)

3.12
監査依頼者（audit client）

監査（**3.1**）を要請する組織又は個人．

注記　内部監査の場合，監査依頼者は，**被監査者**（**3.13**）又は監査プログラムをマネジメントする人でもあり得る．外部監査の要請は，規制当局，契約当事者，潜在的な依頼者又は既存の依頼者からあり得

or existing clients.

[SOURCE: ISO 9000:2015, 3.13.11, modified — Note 1 to entry has been added]

3.13

auditee

organization as a whole or parts thereof being audited

[SOURCE: ISO 9000:2015, 3.13.12, modified]

3.14

audit team

one or more persons conducting an *audit* (**3.1**), supported if needed by *technical experts* (**3.16**)

Note 1 to entry: One *auditor* (**3.15**) of the *audit team* (**3.14**) is appointed as the audit team leader.

Note 2 to entry: The audit team can include auditors-in-training.

[SOURCE: ISO 9000:2015, 3.13.14]

る．

(出典：**JIS Q 9000**:2015 の **3.13.11** を変更．**注記を追加した．**)

3.13
被監査者（auditee）
監査される，組織の全体又はその一部．

(出典：**JIS Q 9000**:2015 の **3.13.12** を変更．)

3.14
監査チーム（audit team）
監査（**3.1**）を行う個人又は複数の人．必要な場合は，**技術専門家**（**3.16**）による支援を受ける．

> 注記1　**監査チーム**（**3.14**）の中の一人の**監査員**（**3.15**）は，監査チームリーダーに指名される．
> 注記2　監査チームには，訓練中の監査員を含めることができる．

(出典：**JIS Q 9000**:2015 の **3.13.14**)

3.15

auditor

person who conducts an *audit* (**3.1**)

[SOURCE: ISO 9000:2015, 3.13.15]

3.16

technical expert

<audit> person who provides specific knowledge or expertise to the *audit team* (**3.14**)

Note 1 to entry: Specific knowledge or expertise relates to the organization, the activity, process, product, service, discipline to be audited, or language or culture.

Note 2 to entry: A technical expert to the *audit team* (**3.14**) does not act as an *auditor* (**3.15**).

[SOURCE: ISO 9000:2015, 3.13.16, modified — Notes to entry 1 and 2 have been modified]

3.17

3.15
監査員(auditor)

　監査(**3.1**)を行う人.

(出典:**JIS Q 9000**:2015 の **3.13.15**)

3.16
技術専門家(technical expert)

　<監査>**監査チーム**(**3.14**)に特定の知識又は専門的技術を提供する人.

　　注記1　特定の知識又は専門的技術とは,監査される組織,活動,プロセス,製品,サービス若しくは監査する分野に関係するもの,又は言語若しくは文化に関係するものである.
　　注記2　**監査チーム**(**3.14**)に対する技術専門家は,**監査員**(**3.15**)としての行動はしない.

(出典:**JIS Q 9000**:2015 の **3.13.16** を変更.**注記1**及び**注記2**を変更した.)

3.17

observer

individual who accompanies the *audit team* (**3.14**) but does not act as an *auditor* (**3.15**)

[SOURCE: ISO 9000:2015, 3.13.17, modified]

3.18

management system

set of interrelated or interacting elements of an organization to establish policies and objectives, and *processes* (**3.24**) to achieve those objectives

Note 1 to entry: A management system can address a single discipline or several disciplines, e.g. quality management, financial management or environmental management.

Note 2 to entry: The management system elements establish the organization's structure, roles and responsibilities, planning, operation, policies, practices, rules, beliefs, objectives and processes to achieve those objectives.

3 用語及び定義

オブザーバ(observer)

監査チーム(3.14)に同行するが,監査員(3.15)として行動しない人.

(出典:**JIS Q 9000**:2015 の **3.13.17** を変更.)

3.18
マネジメントシステム(management system)

方針及び目的(又は目標),並びにその目的(又は目標)を達成するための**プロセス**(3.24)を確立するための,相互に関連する又は相互に作用する,組織の一連の要素.

- 注記 1　一つのマネジメントシステムは,例えば,品質マネジメント,財務マネジメント,環境マネジメントなど,単一又は複数の分野を取り扱うことができる.
- 注記 2　マネジメントシステムの要素は,目的(又は目標)を達成するための,組織の構造,役割及び責任,計画策定,運用,方針,慣行,規則,信条,目的(又は目標),並びにプロセスを確立する.

Note 3 to entry: The scope of a management system can include the whole of the organization, specific and identified functions of the organization, specific and identified sections of the organization, or one or more functions across a group of organizations.

[SOURCE: ISO 9000:2015, 3.5.3, modified — Note 4 to entry has been deleted]

3.19
risk
effect of uncertainty

Note 1 to entry: An effect is a deviation from the expected – positive or negative.

Note 2 to entry: Uncertainty is the state, even partial, of deficiency of information related to, understanding or knowledge of, an event, its consequence and likelihood.

Note 3 to entry: Risk is often characterized by reference to potential events (as defined in ISO Guide 73:2009, 3.5.1.3) and consequences (as defined in

3　用語及び定義　　　51

　　注記3　マネジメントシステムの適用範囲としては，組織全体，組織内の固有で特定された機能，組織内の固有で特定された部門，複数の組織の集まりを横断する一つ又は複数の機能，などがあり得る．

(出典：**JIS Q 9000**:2015 の **3.5.3** を変更．)

3.19
リスク（risk）
不確かさの影響．

　　注記1　影響とは，期待されていることから，好ましい方向又は好ましくない方向にかい（乖）離することをいう．
　　注記2　不確かさとは，事象，その結果及びその起こりやすさに関する，情報，理解又は知識に，たとえ部分的にでも不備がある状態をいう．
　　注記3　リスクは，起こり得る事象（**JIS Q 0073**:2010 の **3.5.1.3** の定義を参照．）及び結果（**JIS Q 0073**:2010 の **3.6.1.3**

ISO Guide 73:2009, 3.6.1.3), or a combination of these.

Note 4 to entry: Risk is often expressed in terms of a combination of the consequences of an event (including changes in circumstances) and the associated likelihood (as defined in ISO Guide 73:2009, 3.6.1.1) of occurrence.

[SOURCE: ISO 9000:2015, 3.7.9, modified — Notes to entry 5 and 6 have been deleted]

3.20
conformity
fulfilment of a *requirement* (**3.23**)

[SOURCE: ISO 9000:2015, 3.6.11, modified — Note 1 to entry has been deleted]

3.21
nonconformity
non-fulfilment of a *requirement* (**3.23**)

[SOURCE: ISO 9000:2015, 3.6.9, modified — Note

の定義を参照.),又はこれらの組合せについて述べることによって,その特徴を示すことが多い.

注記 4　リスクは,ある事象(その周辺状況の変化を含む.)の結果とその発生の起こりやすさ(**JIS Q 0073**:2010 の **3.6.1.1** の定義を参照.)との組合せとして表現されることが多い.

(出典:**JIS Q 9000**:2015 の **3.7.9** を変更.注記 5 及び注記 6 を削除した.)

3.20
適合(conformity)

　要求事項(**3.23**)を満たしていること.

(出典:**JIS Q 9000**:2015 の **3.6.11** を変更.注記 1 及び注記 2 を削除した.)

3.21
不適合(nonconformity)

　要求事項(**3.23**)を満たしていないこと.

(出典:**JIS Q 9000**:2015 の **3.6.9** を変更.注記

1 to entry has been deleted]

3.22
competence

ability to apply knowledge and skills to achieve intended results

[SOURCE: ISO 9000:2015, 3.10.4, modified — Notes to entry have been deleted]

3.23
requirement

need or expectation that is stated, generally implied or obligatory

Note 1 to entry: "Generally implied" means that it is custom or common practice for the organization and interested parties that the need or expectation under consideration is implied.

Note 2 to entry: A specified requirement is one that is stated, for example in documented information.

を削除した．）

3.22
力量（competence）

意図した結果を達成するために，知識及び技能を適用する能力．

（出典：**JIS Q 9000**:2015 の **3.10.4** を変更．**注記 1** 及び **注記 2** を削除した．）

3.23
要求事項（requirement）

明示されている，通常暗黙のうちに了解されている又は義務として要求されている，ニーズ又は期待．

注記 1　"通常暗黙のうちに了解されている" とは，対象となるニーズ又は期待が暗黙のうちに了解されていることが，組織及び利害関係者にとって慣習又は慣行であることを意味する．

注記 2　規定要求事項とは，例えば，文書化した情報の中で明示されている要求事項をいう．

[SOURCE: ISO 9000:2015, 3.6.4, modified — Notes to entry 3, 4, 5 and 6 have been deleted]

3.24

process

set of interrelated or interacting activities that use inputs to deliver an intended result

[SOURCE: ISO 9000:2015, 3.4.1, modified — Notes to entry have been deleted]

3.25

performance

measurable result

Note 1 to entry: Performance can relate either to quantitative or qualitative findings.

Note 2 to entry: Performance can relate to the management of activities, *processes* (**3.24**), products, services, systems or organizations.

[SOURCE: ISO 9000:2015, 3.7.8, modified — Note 3 to entry has been deleted]

（出典：**JIS Q 9000**:2015 の **3.6.4** を変更．**注記 3**〜**注記 6** を削除した．）

3.24
プロセス（process）

インプットを使用して意図した結果を生み出す，相互に関連する又は相互に作用する一連の活動．

（出典：**JIS Q 9000**:2015 の **3.4.1** を変更．**注記 1**〜**注記 6** を削除した．）

3.25
パフォーマンス（performance）

測定可能な結果．

> **注記 1** パフォーマンスは，定量的又は定性的な所見のいずれにも関連し得る．
> **注記 2** パフォーマンスは，活動，**プロセス**（**3.24**），製品，サービス，システム，又は組織の運営管理に関連し得る．

（出典：**JIS Q 9000**:2015 の **3.7.8** を変更．**注記 3** を削除した．）

3.26

effectiveness

extent to which planned activities are realized and planned results achieved

[SOURCE: ISO 9000:2015, 3.7.11, modified — Note 1 to entry has been deleted]

4 Principles of auditing

Auditing is characterized by reliance on a number of principles. These principles should help to make the audit an effective and reliable tool in support of management policies and controls, by providing information on which an organization can act in order to improve its performance. Adherence to these principles is a prerequisite for providing audit conclusions that are relevant and sufficient, and for enabling auditors, working independently from one another, to reach similar conclusions in similar circumstances.

The guidance given in **Clauses 5** to **7** is based on the seven principles outlined below.

a) Integrity: the foundation of professionalism

3.26
有効性(effectiveness)
計画した活動を実行し,計画した結果を達成した程度.

(出典:**JIS Q 9000**:2015 の **3.7.11** を変更.**注記**を削除した.)

4 監査の原則

監査は幾つかの原則に準拠しているという特徴がある.これらの原則は,組織がそのパフォーマンス改善のために行動できる情報を監査が提供することによって,マネジメントの方針及び管理業務を支援する有効な,かつ,信頼のおけるツールとなるのを支援することが望ましい.適切で,かつ,十分な監査結論を導き出すため,そして,互いに独立して監査を行ったとしても同じような状況に置かれれば,どの監査員も同じような結論に達することができるようにするためには,これらの原則の順守は,必須条件である.

この規格の箇条5〜箇条7で示す手引は,次に概要を示す七つの原則に基づく.
a) 高潔さ:専門家であることの基礎

Auditors and the individual(s) managing an audit programme should:

- perform their work ethically, with honesty and responsibility;
- only undertake audit activities if competent to do so;
- perform their work in an impartial manner, i.e. remain fair and unbiased in all their dealings;
- be sensitive to any influences that may be exerted on their judgement while carrying out an audit.

b) Fair presentation: the obligation to report truthfully and accurately

Audit findings, audit conclusions and audit reports should reflect truthfully and accurately the audit activities. Significant obstacles encountered during the audit and unresolved diverging opinions between the audit team and the auditee should be reported. The communication should be truthful, accurate, objective, timely, clear and complete.

c) Due professional care: the application of diligence and judgement in auditing

4 監査の原則

監査員,及び監査プログラムをマネジメントする人は,次の事項を行うことが望ましい.

— 自身の業務を倫理的に,正直に,かつ責任感をもって行う.

— 監査活動を,それを行う力量がある場合にだけ実施する.

— 自身の業務を,公平な進め方で,すなわち,全ての対応において公正さをもち,偏りなく行う.

— 監査の実施中にもたらされるかもしれない,自身の判断へのいかなる影響に対しても,敏感である.

b) 公正な報告:ありのままに,かつ,正確に報告する義務

監査所見,監査結論及び監査報告は,ありのままに,かつ,正確に監査活動を反映することが望ましい.監査中に遭遇した顕著な障害,及び監査チームと被監査者との間で解決に至らない意見の相違について報告することが望ましい.コミュニケーションはありのままに,正確で,客観的で,時宜を得て,明確かつ完全であることが望ましい.

c) 専門家としての正当な注意:監査の際の広範な注意及び判断

Auditors should exercise due care in accordance with the importance of the task they perform and the confidence placed in them by the audit client and other interested parties. An important factor in carrying out their work with due professional care is having the ability to make reasoned judgements in all audit situations.

d) Confidentiality: security of information

Auditors should exercise discretion in the use and protection of information acquired in the course of their duties. Audit information should not be used inappropriately for personal gain by the auditor or the audit client, or in a manner detrimental to the legitimate interests of the auditee. This concept includes the proper handling of sensitive or confidential information.

e) Independence: the basis for the impartiality of the audit and objectivity of the audit conclusions

Auditors should be independent of the activity being audited wherever practicable, and should in all cases act in a manner that is free

監査員は，自らが行っている業務の重要性，並びに監査依頼者及びその他の利害関係者が監査員に対して抱いている信頼に見合う正当な注意を払うことが望ましい．専門家としての正当な注意をもって業務を行う場合の重要な点は，全ての監査状況において根拠ある判断を行う能力をもつことである．

d) 機密保持：情報のセキュリティ

　監査員は，その任務において得た情報の利用及び保護について慎重であることが望ましい．監査情報は，個人的利益のために，監査員又は監査依頼者によって不適切に，又は，被監査者の正当な利益に害をもたらす方法で使用しないことが望ましい．この概念には，取扱いに注意を要する又は機密性のある情報の適切な取扱いを含む．

e) 独立性：監査の公平性及び監査結論の客観性の基礎

　監査員は，実行可能な限り監査の対象となる活動から独立した立場にあり，全ての場合において偏り及び利害抵触がない形で行動すること

from bias and conflict of interest. For internal audits, auditors should be independent from the function being audited if practicable. Auditors should maintain objectivity throughout the audit process to ensure that the audit findings and conclusions are based only on the audit evidence.

For small organizations, it may not be possible for internal auditors to be fully independent of the activity being audited, but every effort should be made to remove bias and encourage objectivity.

f) Evidence-based approach: the rational method for reaching reliable and reproducible audit conclusions in a systematic audit process

Audit evidence should be verifiable. It should in general be based on samples of the information available, since an audit is conducted during a finite period of time and with finite resources. An appropriate use of sampling should be applied, since this is closely related to the confidence that can be placed in the audit conclusions.

g) Risk-based approach: an audit approach that

が望ましい．内部監査では，監査員は，実行可能な場合には，監査の対象となる機能から独立した立場にあることが望ましい．監査員は，監査所見及び監査結論が監査証拠だけに基づくことを確実にするために，監査プロセス中，終始一貫して客観性を維持することが望ましい．

　小規模の組織においては，内部監査員が監査の対象となる活動から完全に独立していることは可能でない場合もあるが，偏りをなくし，客観性を保つあらゆる努力を行うことが望ましい．

f) 証拠に基づくアプローチ：体系的な監査プロセスにおいて，信頼性及び再現性のある監査結論に到達するための合理的な方法

　監査証拠は，検証可能なものであることが望ましい．監査は限られた時間及び資源で行われるので，監査証拠は，一般的に，入手可能な情報からのサンプルに基づくことが望ましい．監査結論にどれだけの信頼をおけるかということと密接に関係しているため，サンプリングを適切に活用することが望ましい．

g) リスクに基づくアプローチ：リスク及び機会を

considers risks and opportunities

The risk-based approach should substantively influence the planning, conducting and reporting of audits in order to ensure that audits are focused on matters that are significant for the audit client, and for achieving the audit programme objectives.

5 Managing an audit programme
5.1 General

An audit programme should be established which can include audits addressing one or more management system standards or other requirements, conducted either separately or in combination (combined audit).

The extent of an audit programme should be based on the size and nature of the auditee, as well as on the nature, functionality, complexity, the type of risks and opportunities, and the level of maturity of the management system(s) to be audited.

The functionality of the management system can be even more complex when most of the important

考慮する監査アプローチ

リスクに基づくアプローチは,監査が,監査依頼者にとって,また,監査プログラムの目的を達成するために重要な事項に焦点を当てることを確実にするため,監査の計画,実施及び報告に対して実質的に影響を及ぼすことが望ましい.

5 監査プログラムのマネジメント
5.1 一般

監査プログラムは,一つ若しくは複数のマネジメントシステム規格又はその他の要求事項に対処し,単独で又は組み合わせて(複合監査)行う監査を含み得るものを確立することが望ましい.

監査プログラムの及ぶ領域は,被監査者の規模及び性質のほか,監査の対象となるマネジメントシステムの性質,機能性,複雑さ,リスク及び機会のタイプ,並びに成熟度に基づくことが望ましい.

マネジメントシステムの機能性は,重要機能の大半を外部委託し,他の組織のリーダーシップの下で

functions are outsourced and managed under the leadership of other organizations. Particular attention needs to be paid to where the most important decisions are made and what constitutes the top management of the management system.

In the case of multiple locations/sites (e.g. different countries), or where important functions are outsourced and managed under the leadership of another organization, particular attention should be paid to the design, planning and validation of the audit programme.

In the case of smaller or less complex organizations the audit programme can be scaled appropriately.

In order to understand the context of the auditee, the audit programme should take into account the auditee's:
— organizational objectives;
— relevant external and internal issues;
— the needs and expectations of relevant interested parties;

マネジメントする場合,更に複雑なものとなり得る.最も重要な決定をどこで下すか,及びマネジメントシステムのトップマネジメントがどのような構成かについて,特別の注意を払う必要がある.

複数の場所・現地(例えば,異なる国々)の場合,又は重要な機能を外部委託し,別の組織のリーダーシップの下でマネジメントする場合,監査プログラムの設計,計画及び妥当性確認に特別の注意を払うことが望ましい.

小規模の又はそれほど複雑でない組織の場合には,監査プログラムの規模は,適切に縮小できる.

被監査者の状況を理解するために,監査プログラムは,被監査者について,次の事項を考慮に入れることが望ましい.
— 組織の目的
— 関連する外部及び内部の課題
— 関連する利害関係者のニーズ及び期待

— information security and confidentiality requirements.

The planning of internal audit programmes and, in some cases programmes for auditing external providers, can be arranged to contribute to other objectives of the organization.

The individual(s) managing the audit programme should ensure the integrity of the audit is maintained and that there is not undue influence exerted over the audit.

Audit priority should be given to allocating resources and methods to matters in a management system with higher inherent risk and lower level of performance.

Competent individuals should be assigned to manage the audit programme.

The audit programme should include information and identify resources to enable the audits to be conducted effectively and efficiently within the

— 情報セキュリティ及び機密保持の要求事項

　内部監査プログラムの計画の策定，及び場合によっては外部提供者を監査するプログラムの計画の策定は，組織の他の目的にも寄与するように取り決めることができる．

　監査プログラムをマネジメントする人は，監査の"完全に整っている状態"（integrity）を維持し，監査に過度の影響が及ばないことを確実にすることが望ましい．

　監査の優先順位は，マネジメントシステムにおいて内在するリスクが高く，パフォーマンスレベルが低い事項に対して資源及び手法を割り当てるよう，与えられることが望ましい．

　監査プログラムをマネジメントするためには，力量のある人を割り当てることが望ましい．

　監査プログラムは，決められた期間内で有効にかつ効率的に監査を行えるようにするための情報を含めて，資源を特定することが望ましい．そのような

specified time frames. The information should include:

a) objectives for the audit programme;

b) risks and opportunities associated with the audit programme (see **5.3**) and the actions to address them;

c) scope (extent, boundaries, locations) of each audit within the audit programme;

d) schedule (number/duration/frequency) of the audits;

e) audit types, such as internal or external;

f) audit criteria;

g) audit methods to be employed;

h) criteria for selecting audit team members;

i) relevant documented information.

Some of this information may not be available until more detailed audit planning is complete.

The implementation of the audit programme should be monitored and measured on an ongoing basis (see **5.6**) to ensure its objectives have been achieved. The audit programme should be reviewed in order to identify needs for changes and

5 監査プログラムのマネジメント

情報には，次の事項を含めることが望ましい．

a) 監査プログラムの目的
b) 監査プログラムに付随するリスク及び機会(**5.3** 参照）並びにそれらに対処する活動

c) 監査プログラム内の各監査の範囲（及ぶ領域，境界及び場所）
d) 監査のスケジュール（回数・期間・頻度）

e) 監査のタイプ，例えば内部監査又は外部監査
f) 監査基準
g) 採用する監査方法
h) 監査チームメンバーの選定基準
i) 関連する文書化した情報

これらの情報の幾つかは，より詳細な監査計画の策定が完了するまでは利用できない場合がある．

監査プログラムの実施状況を，その目的が達成されていることを確実にするために継続的に監視し，測定する（**5.6** 参照）ことが望ましい．監査プログラムは，変更の必要性及び改善の機会の可能性を特定するためにレビューすることが望ましい（**5.7** 参

possible opportunities for improvements (see **5.7**).

Figure 1 illustrates the process flow for the management of an audit programme.

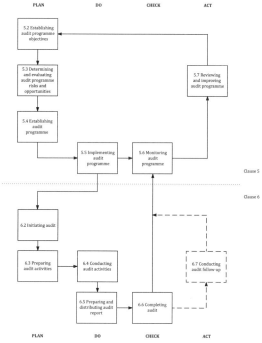

NOTE 1   This Figure illustrates the application of the Plan-Do-Check-Act cycle in this document.

NOTE 2   Clause/subclause numbering refers to the relevant clauses/subclauses of this document.

Figure 1 — Process flow for the management of an audit programme

照).

監査プログラムのマネジメントのためのプロセスフローを**図1**に示す．

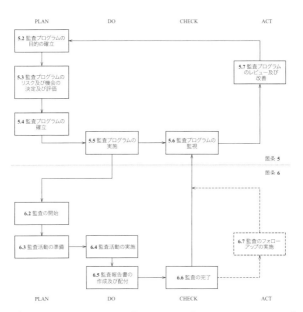

注記1 この図は，この規格における"Plan-Do-Check-Act"（PDCA）サイクルの適用について示している．
注記2 箇条・細分箇条の番号付けは，この規格の関連する箇条・細分箇条番号を示す．

図1−監査プログラムのマネジメントのためのプロセスフロー

5.2 Establishing audit programme objectives

The audit client should ensure that the audit programme objectives are established to direct the planning and conducting of audits and should ensure the audit programme is implemented effectively. Audit programme objectives should be consistent with the audit client's strategic direction and support management system policy and objectives.

These objectives can be based on consideration of the following:

a) needs and expectations of relevant interested parties, both external and internal;

b) characteristics of and requirements for processes, products, services and projects, and any changes to them;

c) management system requirements;

d) need for evaluation of external providers;

e) auditee's level of performance and level of maturity of the management system(s), as reflected in relevant performance indicators (e.g. KPIs), the occurrence of nonconformities

5.2 監査プログラムの目的の確立

　監査依頼者は，監査の計画策定及び実施を指示するために監査プログラムの目的が確立され，その監査プログラムが有効に実施されるのを確実にすることが望ましい．監査プログラムの目的は，監査依頼者の戦略的方向と整合し，マネジメントシステムの方針及び目的（又は目標）を支持するものであることが望ましい．

　監査プログラムの目的は，次の考慮事項に基づき得る．

a) 外部及び内部双方の，関連する利害関係者のニーズ及び期待
b) プロセス，製品，サービス及びプロジェクトの特性並びにそれらに関わる要求事項，並びにそれらに対する変化
c) マネジメントシステムの要求事項
d) 外部提供者を評価することの必要性
e) 被監査者のマネジメントシステムに関する，パフォーマンスのレベル及び成熟度．それらは，関連するパフォーマンス指標（例：**KPI**），不適合若しくはインシデントの発生，又は利害関

or incidents or complaints from interested parties;
f) identified risks and opportunities to the auditee;
g) results of previous audits.

Examples of audit programme objectives can include the following:
— identify opportunities for the improvement of a management system and its performance;
— evaluate the capability of the auditee to determine its context;
— evaluate the capability of the auditee to determine risks and opportunities and to identify and implement effective actions to address them;
— conform to all relevant requirements, e.g. statutory and regulatory requirements, compliance commitments, requirements for certification to a management system standard;
— obtain and maintain confidence in the capability of an external provider;
— determine the continuing suitability, adequacy and effectiveness of the auditee's manage-

係者からの苦情が反映されたものである．

f) 被監査者に対して特定されたリスク及び機会

g) 前回までの監査の結果

　監査プログラムの目的の例には，次の事項を含み得る．
— マネジメントシステム及びそのパフォーマンスの改善の機会を特定する．
— 被監査者が自身の状況を明確にする能力を評価する．
— リスク及び機会を決定し，それらに対処する有効な活動を特定し実施する，被監査者の能力を評価する．

— 全ての関連する要求事項，例えば法令・規制要求事項，順守のコミットメント，マネジメントシステム規格の認証に関する要求事項に適合する．
— 外部提供者の能力における信頼を獲得し，維持する．
— 被監査者のマネジメントシステムの継続的な適切性，妥当性及び有効性を決定する．

ment system;

— evaluate the compatibility and alignment of the management system objectives with the strategic direction of the organization.

5.3 Determining and evaluating audit programme risks and opportunities

There are risks and opportunities related to the context of the auditee that can be associated with an audit programme and can affect the achievement of its objectives. The individual(s) managing the audit programme should identify and present to the audit client the risks and opportunities considered when developing the audit programme and resource requirements, so that they can be addressed appropriately.

There can be risks associated with the following:

a) planning, e.g. failure to set relevant audit objectives and determine the extent, number, duration, locations and schedule of the audits;

b) resources, e.g. allowing insufficient time, equipment and/or training for developing the audit programme or conducting an audit;

― マネジメントシステムの目的（又は目標）が，組織の戦略的方向と両立し，整合しているかを評価する．

5.3 監査プログラムのリスク及び機会の決定及び評価

　被監査者の状況に関係してリスク及び機会があり，それらは監査プログラムに付随し，その目的の達成に影響を及ぼし得る．監査プログラムをマネジメントする人は，監査プログラム及び資源に関する要求事項を策定する際に考慮されるリスク及び機会に適切に対処するために，それらを特定し，監査依頼者に対して提示することが望ましい．

　次の事項に付随するリスクがあり得る．
a) 計画の策定．例えば，関連する監査目的の設定における失敗，並びに監査の及ぶ領域，回数，期間，場所及びスケジュールの決定における失敗．
b) 資源．例えば，監査プログラムの策定又は監査の実施に十分な時間，機器及び／又は訓練を与えない．

c) selection of the audit team, e.g. insufficient overall competence to conduct audits effectively;

d) communication, e.g. ineffective external/internal communication processes/channels;

e) implementation, e.g. ineffective coordination of the audits within the audit programme, or not considering information security and confidentiality;

f) control of documented information, e.g. ineffective determination of the necessary documented information required by auditors and relevant interested parties, failure to adequately protect audit records to demonstrate audit programme effectiveness;

g) monitoring, reviewing and improving the audit programme, e.g. ineffective monitoring of audit programme outcomes;

h) availability and cooperation of auditee and availability of evidence to be sampled.

Opportunities for improving the audit programme can include:

5 監査プログラムのマネジメント

c) 監査チームの選定．例えば，監査を有効に行う全体としての力量が不十分である．

d) コミュニケーション．例えば，外部・内部コミュニケーションのプロセス・手段が有効でない．

e) 実施．例えば，監査プログラム内における調整が有効でない，又は情報セキュリティ及び機密保持を考慮していない．

f) 文書化した情報の管理．例えば，監査員及び関連する利害関係者が必要とする文書化した情報の決定が有効でなく，監査プログラムの有効性を実証するための監査記録の保護が十分でない．

g) 監査プログラムの監視，レビュー及び改善．例えば，監査プログラムの成果が有効に監視されていない．

h) 被監査者の参加可能性及び協力，並びにサンプリングする証拠の利用可能性

監査プログラムを改善する機会には，次の事項を含み得る．

- allowing multiple audits to be conducted in a single visit;
- minimizing time and distances travelling to site;
- matching the level of competence of the audit team to the level of competence needed to achieve the audit objectives;
- aligning audit dates with the availability of auditee's key staff.

5.4 Establishing the audit programme

5.4.1 Roles and responsibilities of the individual(s) managing the audit programme

The individual(s) managing the audit programme should:

a) establish the extent of the audit programme according to the relevant objectives (see **5.2**) and any known constraints;

b) determine the external and internal issues, and risks and opportunities that can affect the audit programme, and implement actions to address them, integrating these actions in all relevant auditing activities, as appropriate;

― 一回の訪問で複数の監査を行うことを認める．

― 現地への移動時間及び距離を最小限にする．

― 監査チームの力量レベルを，監査目的を達成するために必要な力量レベルに合わせる．

― 監査日を，被監査者の主要なスタッフが参加可能な日に合わせる．

5.4 監査プログラムの確立
5.4.1 監査プログラムをマネジメントする人の役割及び責任

監査プログラムをマネジメントする人は，次の事項を行うことが望ましい．

a) 監査プログラムの及ぶ領域を，関連のある目的（**5.2参照**）及び全ての既知の制約に基づいて確立する．

b) 外部及び内部の課題，並びに監査プログラムに影響を及ぼし得るリスク及び機会を決定し，それらに対処する活動を実施し，必要に応じて，これらの活動を全ての関連する監査活動に統合する．

c) ensuring the selection of audit teams and the overall competence for the auditing activities by assigning roles, responsibilities and authorities, and supporting leadership, as appropriate;

d) establish all relevant processes including processes for:
   — the coordination and scheduling of all audits within the audit programme;
   — the establishment of audit objectives, scope(s) and criteria of the audits, determining audit methods and selecting the audit team;
   — evaluating auditors;
   — the establishment of external and internal communication processes, as appropriate;
   — the resolutions of disputes and handling of complaints;
   — audit follow-up if applicable;
   — reporting to the audit client and relevant interested parties, as appropriate.

e) determine and ensure provision of all necessary resources;

5 監査プログラムのマネジメント

c) 監査チームの選定及び監査活動についての全体としての力量を確実にする．これは，役割，責任及び権限を割り当て，必要に応じて，リーダーシップを支援することによる．

d) 次の事項のためのプロセスを含む，全ての関連プロセスを確立する．
   — 監査プログラム内の全ての監査の調整及びスケジュールの策定
   — 監査の目的，範囲及び基準の確立，監査方法の決定，並びに監査チームの選定

   — 監査員の評価
   — 必要に応じて，外部及び内部コミュニケーションプロセスの確立

   — 紛争の解決及び苦情の取扱い

   — 該当する場合は，監査のフォローアップ
   — 必要に応じて，監査依頼者及び関連する利害関係者への報告

e) 全ての必要な資源の提供を決定し，それを確実にする．

f) ensure that appropriate documented information is prepared and maintained, including audit programme records;

g) monitor, review and improve the audit programme;

h) communicate the audit programme to the audit client and, as appropriate, relevant interested parties.

The individual(s) managing the audit programme should request its approval by the audit client.

5.4.2 Competence of individual(s) managing audit programme

The individual(s) managing the audit programme should have the necessary competence to manage the programme and its associated risks and opportunities and external and internal issues effectively and efficiently, including knowledge of:

a) audit principles (see **Clause 4**), methods and processes (see **A.1** and **A.2**);

b) management system standards, other relevant standards and reference/guidance docu-

5 監査プログラムのマネジメント 89

- **f)** 適切な文書化した情報を作成し維持することを確実にする.これには,監査プログラムの記録を含む.
- **g)** 監査プログラムを監視し,レビューし,改善する.
- **h)** 監査プログラムを監査依頼者,及び必要に応じて,関連する利害関係者へ伝達する.

　監査プログラムをマネジメントする人は,監査依頼者に監査プログラムの承認を要請することが望ましい.

5.4.2　監査プログラムをマネジメントする人の力量

　監査プログラムをマネジメントする人は,監査プログラム及びそれに付随するリスク及び機会,並びに外部及び内部の課題を有効にかつ効率的にマネジメントするために必要な力量を備えていることが望ましい.これには,次の事項に関する知識を含む.

- **a)** 監査の原則(箇条 **4** 参照),方法及びプロセス(**A.1** 及び **A.2** 参照)
- **b)** マネジメントシステム規格,その他の関連する規格及び基準・手引文書

ments;

c) information regarding the auditee and its context (e.g. external/internal issues, relevant interested parties and their needs and expectations, business activities, products, services and processes of the auditee);

d) applicable statutory and regulatory requirements and other requirements relevant to the business activities of the auditee.

As appropriate, knowledge of risk management, project and process management, and information and communications technology (ICT) may be considered.

The individual(s) managing the audit programme should engage in appropriate continual development activities to maintain the necessary competence to manage the audit programme.

5.4.3 Establishing extent of audit programme

The individual(s) managing the audit programme should determine the extent of the audit pro-

c) 被監査者及びその状況に関わる情報（例えば，外部・内部の課題，関連する利害関係者並びにそのニーズ及び期待，被監査者の事業活動，製品，サービス及びプロセス）

d) 適用される法令・規制の要求事項，及び被監査者の事業活動に関連するその他の要求事項

　必要な場合には，リスクマネジメント，プロジェクト及びプロセスのマネジメント，並びに情報通信技術（ICT）に関する知識を考慮してよい．

　監査プログラムをマネジメントする人は，監査プログラムをマネジメントするのに必要な力量を維持するために，適切な継続的開発活動に携わることが望ましい．

5.4.3　監査プログラムの及ぶ領域の確立

　監査プログラムをマネジメントする人は，監査プログラムの及ぶ領域を決定することが望ましい．監

gramme. This can vary depending on the information provided by the auditee regarding its context (see **5.3**).

NOTE   In certain cases, depending on the auditee's structure or its activities, the audit programme might only consist of a single audit (e.g. a small project or organization).

Other factors impacting the extent of an audit programme can include the following:

a) the objective, scope and duration of each audit and the number of audits to be conducted, reporting method and, if applicable, audit follow up;

b) the management system standards or other applicable criteria;

c) the number, importance, complexity, similarity and locations of the activities to be audited;

d) those factors influencing the effectiveness of the management system;

e) applicable audit criteria, such as planned arrangements for the relevant management system standards, statutory and regulatory re-

査プログラムの及ぶ領域は，被監査者が自身の状況に関して提供する情報によって異なり得る（**5.3**参照）．

> **注記** 被監査者の組織構造又は活動によって，監査プログラムは単一の監査だけから成る場合もある（例えば，小さなプロジェクト又は組織）．

監査プログラムの及ぶ領域に影響を与えるその他の要因には，次の事項を含み得る．

a) 実施するそれぞれの監査の目的，範囲及び期間，並びに監査の実施回数，報告方法，及び該当する場合は，監査のフォローアップ

b) マネジメントシステム規格又はその他の適用可能な基準

c) 監査の対象となる活動の数，重要性，複雑さ，類似性及び場所

d) マネジメントシステムの有効性に影響を与える要因

e) 適用される監査基準．例えば，関連するマネジメントシステム規格のために計画された取決め事項，法令・規制要求事項並びに被監査者であ

quirements and other requirements to which the organization is committed;

f) results of previous internal or external audits and management reviews, if appropriate;

g) results of a previous audit programme review;

h) language, cultural and social issues;

i) the concerns of interested parties, such as customer complaints, non-compliance with statutory and regulatory requirements and other requirements to which the organization is committed, or supply chain issues;

j) significant changes to the auditee's context or its operations and related risks and opportunities;

k) availability of information and communication technologies to support audit activities, in particular the use of remote audit methods (see **A.16**);

l) the occurrence of internal and external events, such as nonconformities of products or service, information security leaks, health and safety incidents, criminal acts or environmental incidents;

m) business risks and opportunities, including

る組織がコミットメントしたその他の要求事項

f) 前回までの内部監査又は外部監査の結果,及び該当する場合は,マネジメントレビューの結果
g) 前回の監査プログラムのレビュー結果
h) 言語,文化及び社会上の課題
i) 利害関係者の懸念事項.例えば,顧客の苦情,法令・規制要求事項及び被監査者である組織がコミットメントしたその他の要求事項への不順守,又はサプライチェーンの課題

j) 被監査者の状況又はその運用並びに関連するリスク及び機会に対する重大な変化

k) 監査活動を支援する,被監査者の情報通信技術の利用可能性.特に遠隔監査方法の利用(**A.16**参照)

l) 内部及び外部の事象の発生.例えば,製品又はサービスの不適合,情報セキュリティ漏えい(洩),安全衛生に関わるインシデント,犯罪行為又は環境に関わるインシデントなど.

m) 事業のリスク及び機会.これには,それらに対

actions to address them.

5.4.4 Determining audit programme resources

When determining resources for the audit programme, the individual(s) managing the audit programme should consider:

a) the financial and time resources necessary to develop, implement, manage and improve audit activities;

b) audit methods (see **A.1**);

c) the individual and overall availability of auditors and technical experts having competence appropriate to the particular audit programme objectives;

d) the extent of the audit programme (see **5.4.3**) and audit programme risks and opportunities (see **5.3**);

e) travel time and cost, accommodation and other auditing needs;

f) the impact of different time zones;

g) the availability of information and communication technologies (e.g. technical resources required to set up a remote audit using tech-

処する活動を含む.

5.4.4 監査プログラムの資源の決定

監査プログラムをマネジメントする人は,監査プログラムの資源の決定に当たって,次の事項を考慮することが望ましい.

a) 監査活動を計画し,実施し,マネジメントし,改善するために必要な財務資源及び工数

b) 監査方法(**A.1** 参照)

c) 特定の監査プログラムの目的にふさわしい力量を備えた,監査員及び技術専門家の個別の及び全体的な利用可能性

d) 監査プログラムの及ぶ領域(**5.4.3** 参照)並びに監査プログラムのリスク及び機会(**5.3** 参照)

e) 移動時間及び費用,宿泊施設並びにその他監査に必要な事項

f) 異なったタイムゾーンの影響

g) 情報通信技術の利用可能性(例えば,遠隔の協調活動を支援する技術を用いた,遠隔監査を設定するために必要な技術資源)

nologies that support remote collaboration);

h) the availability of any tools, technology and equipment required;

i) the availability of necessary documented information, as determined during the establishment of the audit programme (see **A.5**);

j) requirements related to the facility, including any security clearances and equipment (e.g. background checks, personal protective equipment, ability to wear clean room attire).

5.5 Implementing audit programme
5.5.1 General

Once the audit programme has been established (see **5.4.3**) and related resources have been determined (see **5.4.4**) it is necessary to implement the operational planning and the coordination of all the activities within the programme.

The individual(s) managing the audit programme should:

a) communicate the relevant parts of the audit programme, including the risks and opportunities involved, to relevant interested parties

h) 必要なあらゆるツール,技術及び機器の利用可能性
i) 監査プログラムの確立において決定した,必要な文書化した情報の利用可能性(**A.5** 参照)

j) 施設に関する要求事項.これには,あらゆる機密情報取扱い許可及び機器を含む(例えば,身元調査,個人用保護具,クリーンルーム用着衣を着用する能力).

5.5 監査プログラムの実施
5.5.1 一般
　監査プログラムを確立し(**5.4.3** 参照)関係する資源を決定したなら(**5.4.4** 参照),その運用計画の策定及び監査プログラム内の全ての活動の調整を実施する必要がある.

　監査プログラムをマネジメントする人は,次の事項を行うことが望ましい.
a) 関係するリスク及び機会を含め,監査プログラムの関連する部分を関連する利害関係者に伝達する,並びに確立した外部及び内部コミュニケ

and inform them periodically of its progress, using established external and internal communication channels;

b) define objectives, scope and criteria for each individual audit;
c) select audit methods (see **A.1**);
d) coordinate and schedule audits and other activities relevant to the audit programme;
e) ensure the audit teams have the necessary competence (see **5.5.4**);
f) provide necessary individual and overall resources to the audit teams (see **5.4.4**);
g) ensure the conduct of audits in accordance with the audit programme, managing all operational risks, opportunities and issues (i.e. unexpected events), as they arise during the deployment of the programme;
h) ensure relevant documented information regarding the auditing activities is properly managed and maintained (see **5.5.7**);
i) define and implement the operational controls (see **5.6**) necessary for audit programme monitoring;

5　監査プログラムのマネジメント

ーションチャネルを用いて，関連する利害関係者に定期的にその進捗状況を知らせる．

b) 個々の監査の，目的，範囲及び基準を定める．

c) 監査方法を選択する（**A.1** 参照）．

d) 監査プログラムに関連する，監査及びその他の活動について，調整及びスケジュールの作成をする．

e) 監査チームが必要な力量をもつことを確実にする（**5.5.4** 参照）．

f) 監査チームに，必要な個別の資源及び全体的な資源を提供する（**5.4.4** 参照）．

g) 監査プログラムに従った監査を行うことを確実にする．それは，監査プログラムの実施展開中に発生する，全ての運用上のリスク，機会，及び課題（すなわち，予期しない事象）をマネジメントすることである．

h) 監査活動に関わる文書化した情報を，適切にマネジメントし，維持することを確実にする（**5.5.7** 参照）．

i) 監査プログラムの監視に必要な運用上の管理（**5.6** 参照）を定め，実施する．

j) review the audit programme in order to identify opportunities for its improvement (see **5.7**).

5.5.2 Defining the objectives, scope and criteria for an individual audit

Each individual audit should be based on defined audit objectives, scope and criteria. These should be consistent with the overall audit programme objectives.

The audit objectives define what is to be accomplished by the individual audit and may include the following:

a) determination of the extent of conformity of the management system to be audited, or parts of it, with audit criteria;

b) evaluation of the capability of the management system to assist the organization in meeting relevant statutory and regulatory requirements and other requirements to which the organization is committed;

c) evaluation of the effectiveness of the management system in meeting its intended results;

j) 監査プログラムの改善の機会を特定するために，監査プログラムをレビューする（**5.7** 参照）．

5.5.2 個々の監査の目的，範囲及び基準の定義

個々の監査は，定められた監査の目的，範囲及び基準に基づくことが望ましい．これらは，全体的な監査プログラムの目的と整合していることが望ましい．

監査目的は，その個々の監査で何を達成するのかを定めるものであり，次の事項を含めてよい．

a) 監査の対象となるマネジメントシステム又はその一部の，監査基準への適合の程度の決定

b) 関連する法令・規制要求事項及び組織がコミットメントしたその他の要求事項を満たす上で組織を支援する，マネジメントシステムの能力の評価

c) 意図した結果を満たす上でのマネジメントシステムの有効性の評価

d) identification of opportunities for potential improvement of the management system;

e) evaluation of the suitability and adequacy of the management system with respect to the context and strategic direction of the auditee;

f) evaluation of the capability of the management system to establish and achieve objectives and effectively address risks and opportunities, in a changing context, including the implementation of the related actions.

The audit scope should be consistent with the audit programme and audit objectives. It includes such factors as locations, functions, activities and processes to be audited, as well as the time period covered by the audit.

The audit criteria are used as a reference against which conformity is determined. These may include one or more of the following: applicable policies, processes, procedures, performance criteria including objectives, statutory and regulatory requirements, management system requirements, information regarding the context and the risks

d) マネジメントシステムの潜在的な改善の機会の特定

e) 被監査者の状況及び戦略的方向性に関する，マネジメントシステムの適切性及び妥当性の評価

f) 目的（又は目標）を確立及び達成し，変化する状況においてリスク及び機会に有効に対処するための，マネジメントシステムの能力の評価．これには，関係する活動の実施を含む．

監査範囲は，監査プログラム及び監査目的と整合していることが望ましい．これには，監査の対象となる，場所，機能，活動及びプロセス，並びに監査期間のような要素を含む．

監査基準は，適合性を決定する基準として用いられる．これには，次の事項の一つ又は複数を含めてよい．適用される方針，プロセス，手順，目的（又は目標）を含むパフォーマンス基準，法令・規制要求事項，マネジメントシステムの要求事項，被監査者が決定した状況並びにリスク及び機会に関する情報（関連する外部・内部利害関係者の要求事項を含

and opportunities as determined by the auditee (including relevant external/internal interested parties requirements), sector codes of conduct or other planned arrangements.

In the event of any changes to the audit objectives, scope or criteria, the audit programme should be modified if necessary and communicated to interested parties, for approval if appropriate.

When more than one discipline is being audited at the same time it is important that the audit objectives, scope and criteria are consistent with the relevant audit programmes for each discipline. Some disciplines can have a scope that reflects the whole organization and others can have a scope that reflects a subset of the whole organization.

5.5.3 Selecting and determining audit methods

The individual(s) managing the audit programme should select and determine the methods for effectively and efficiently conducting an audit, depending on the defined audit objectives, scope and crite-

む．），業界の行動規範又はその他の計画された取決め事項．

　監査の目的，範囲又は基準に何らかの変更があった場合には，必要に応じて監査プログラムを修正し，利害関係者に伝えて，適宜その承認を求めることが望ましい．

　複数の分野を同時に監査する場合，監査の目的，範囲及び基準が，各分野の関連する監査プログラムと整合していることが重要である．組織全体を反映する監査範囲をもち得る分野もあれば，組織のある部分を反映する監査範囲をもち得る分野もある．

5.5.3　監査方法の選択及び決定

　監査プログラムをマネジメントする人は，定められた監査の目的，範囲及び基準に基づいて，監査を有効にかつ効率的に行うための方法を選択し，決定することが望ましい．

ria.

Audits can be performed on-site, remotely or as a combination. The use of these methods should be suitably balanced, based on, among others, consideration of associated risks and opportunities.

Where two or more auditing organizations conduct a joint audit of the same auditee, the individuals managing the different audit programmes should agree on the audit methods and consider implications for resourcing and planning the audit. If an auditee operates two or more management systems of different disciplines, combined audits may be included in the audit programme.

5.5.4 Selecting audit team members

The individual(s) managing the audit programme should appoint the members of the audit team, including the team leader and any technical experts needed for the specific audit.

An audit team should be selected, taking into account the competence needed to achieve the objec-

監査は,現地,遠隔,又はこれらを組み合わせて実施することができる.これらの方法は,とりわけ,付随するリスク及び機会の考慮に基づいて,適切にバランスをとって利用することが望ましい.

複数の監査組織が同一の被監査者に対して合同監査を行う場合は,異なる監査プログラムをマネジメントする人は,監査方法について合意し,監査の資源提供及び監査計画の策定への影響を考慮することが望ましい.被監査者が異なった分野の複数のマネジメントシステムを運用している場合は,監査プログラムには複合監査を含めてよい.

5.5.4 監査チームメンバーの選定

監査プログラムをマネジメントする人は,チームリーダー及び特定の監査に必要な技術専門家を含めて,監査チームメンバーを指名することが望ましい.

監査チームは,定められた監査範囲の中で個々の監査の目的を達成するために必要な力量を考慮に入

tives of the individual audit within the defined scope. If there is only one auditor, the auditor should perform all applicable duties of an audit team leader.

NOTE **Clause 7** contains guidance on determining the competence required for the audit team members and describes the processes for evaluating auditors.

To assure the overall competence of the audit team, the following steps should be performed:
— identification of the competence needed to achieve the objectives of the audit;
— selection of the audit team members so that the necessary competence is present in the audit team.

In deciding the size and composition of the audit team for the specific audit, consideration should be given to the following:
a) the overall competence of the audit team needed to achieve audit objectives, taking into account audit scope and criteria;

れて,選定することが望ましい.監査員が一人だけの場合は,その監査員が監査チームリーダーとしての適用される全ての任務を果たすことが望ましい.

> 注記 箇条7は,監査チームメンバーに求められる力量を決定するための手引を示し,かつ,監査員を評価するプロセスを示している.

監査チームの全体としての力量を保証するために,次のステップを実施することが望ましい.
— 監査の目的を達成するために必要な力量の特定

— 監査チームに必要な力量が存在するような,監査チームメンバーの選定

個別の監査のための監査チームの規模及び構成を決めるに当たって,次の事項を考慮することが望ましい.
a) 監査範囲及び監査基準を考慮に入れた,監査目的を達成するために必要な監査チーム全体としての力量

b) complexity of the audit;

c) whether the audit is a combined or joint audit;

d) the selected audit methods;

e) ensuring objectivity and impartiality to avoid any conflict of interest of the audit process;

f) the ability of the audit team members to work and interact effectively with the representatives of the auditee and relevant interested parties;

g) the relevant external/internal issues, such as the language of the audit, and the auditee's social and cultural characteristics. These issues may be addressed either by the auditor's own skills or through the support of a technical expert, also considering the need for interpreters;

h) type and complexity of the processes to be audited.

Where appropriate, the individual(s) managing the audit programme should consult the team leader on the composition of the audit team.

If the necessary competence is not covered by the

b) 監査の複雑さ
c) 監査が複合監査又は合同監査であるかどうか
d) 選択された監査方法
e) 監査プロセスのあらゆる利害抵触を回避するための,客観性及び公平性の確保
f) 被監査者の代表者及び関連する利害関係者と,有効に作業し相互調整を行うための監査チームメンバーの能力

g) 関連する外部・内部の課題,監査で使用する言語,被監査者の社会的及び文化的特徴など.これらの課題には,監査員自身の技能によって,又は技術専門家による支援を介して対処してもよく,その際,通訳者の必要性も考慮する.

h) 監査対象となるプロセスのタイプ及び複雑さ

必要に応じて,監査プログラムをマネジメントする人は,監査チームの構成について,チームリーダーに意見を求めることが望ましい.

監査チームの監査員だけでは必要な力量が確保で

auditors in the audit team, technical experts with additional competence should be made available to support the team.

Auditors-in-training may be included in the audit team, but should participate under the direction and guidance of an auditor.

Changes to the composition of the audit team may be necessary during the audit, e.g. if a conflict of interest or competence issue arises. If such a situation arises, it should be resolved with the appropriate parties (e.g. audit team leader, the individual(s) managing the audit programme, audit client or auditee) before any changes are made.

5.5.5 Assigning responsibility for an individual audit to the audit team leader

The individual(s) managing the audit programme should assign the responsibility for conducting the individual audit to an audit team leader.

The assignment should be made in sufficient time before the scheduled date of the audit, in order to

きない場合は,追加的な力量を備えた技術専門家が,そのチームを支援するために利用可能であることが望ましい.

訓練中の監査員を監査チームに含めてよいが,監査員の指揮及び指導の下で参加させることが望ましい.

監査中に監査チームの構成の変更が必要となる場合がある.例えば,利害抵触又は力量に関する課題が生じた場合である.このような状況が生じたならば,いかなる変更でもそれを行う前に,適切な関係者(例えば,監査チームリーダー,監査プログラムをマネジメントする人,監査依頼者又は被監査者)とその状況を解決しておくことが望ましい.

5.5.5 監査チームリーダーに対する,個々の監査の責任の割当て

監査プログラムをマネジメントする人は,個々の監査を行う責任を監査チームリーダーに割り当てることが望ましい.

監査の有効な計画策定を確実にするために,この割当ては,監査の計画された期日前に十分な時間を

ensure the effective planning of the audit.

To ensure effective conduct of the individual audits, the following information should be provided to the audit team leader:

a) audit objectives;
b) audit criteria and any relevant documented information;
c) audit scope, including identification of the organization and its functions and processes to be audited;
d) audit processes and associated methods;
e) composition of the audit team;
f) contact details of the auditee, the locations, time frame and duration of the audit activities to be conducted;
g) resources necessary to conduct the audit;
h) information needed for evaluating and addressing identified risks and opportunities to the achievement of the audit objectives;
i) information which supports the audit team leader(s) in their interactions with the auditee for the effectiveness of the audit programme.

もって行うことが望ましい.

　個々の監査を有効に行うことを確実にするために,監査チームリーダーに次の情報を提供することが望ましい.
a) 監査目的
b) 監査基準及びあらゆる関連する文書化した情報

c) 監査範囲.これには,監査の対象となる組織及び機能並びにプロセスの特定を含む.

d) 監査プロセス及びそれに付随する方法
e) 監査チームの構成
f) 被監査者の連絡先,監査活動を行う,場所,期間(time frame),及び工数

g) 監査の実施に必要な資源
h) 監査目的の達成に対する,特定したリスク及び機会の,評価及び対処に必要な情報

i) 監査チームリーダーが監査プログラムの有効性について被監査者とやりとりする際に,監査チームリーダーの支援となる情報

The assignment information should also cover the following, as appropriate:

— working and reporting language of the audit where this is different from the language of the auditor or the auditee, or both;
— audit reporting output as required and to whom it is to be distributed;
— matters related to confidentiality and information security, as required by the audit programme;
— any health, safety and environmental arrangements for the auditors;
— requirements for travel or access to remote sites;
— any security and authorization requirements;
— any actions to be reviewed, e.g. follow-up actions from a previous audit;
— coordination with other audit activities, e.g. when different teams are auditing similar or related processes at different locations or in the case of a joint audit.

Where a joint audit is conducted, it is important to

5 監査プログラムのマネジメント

該当する場合，この割当てに関する情報には，次の事項も網羅することが望ましい．
— 監査の作業及び報告に用いる言語で，監査員若しくは被監査者又はその両方の言語と異なる場合
— 必要な監査報告アウトプット及びその配付先

— 監査プログラムが求めるような，機密保持及び情報セキュリティに関係する事項

— 監査員に対する，安全衛生上及び環境上のあらゆる取決め
— 遠隔サイトへの，移動又はアクセスに関する要求事項
— あらゆるセキュリティ及び権限付与に関する要求事項
— レビューすべきあらゆる活動．例えば，前回の監査からのフォローアップ
— その他の監査活動との調整．例えば，異なるチームが異なる場所で類似の若しくは関係するプロセスを監査している場合，又は合同監査の場合

合同監査を行う場合，監査を開始する前に，監査

reach agreement among the organizations conducting the audits, before the audit commences, on the specific responsibilities of each party, particularly with regard to the authority of the team leader appointed for the audit.

5.5.6 Managing audit programme results

The individual(s) managing the audit programme should ensure that the following activities are performed:

a) evaluation of the achievement of the objectives for each audit within the audit programme;

b) review and approval of audit reports regarding the fulfilment of the audit scope and objectives;

c) review of the effectiveness of actions taken to address audit findings;

d) distribution of audit reports to relevant interested parties;

e) determination of the necessity for any follow-up audit.

The individual managing the audit programme should consider, where appropriate:

を行う組織間でそれぞれの責任,特に監査のために指名されたチームリーダーの権限について,合意に達していることが重要である.

5.5.6 監査プログラムの結果のマネジメント

監査プログラムをマネジメントする人は,次の活動が行われることを確実にすることが望ましい.

a) 監査プログラム内の各監査における,監査目的の達成についての評価
b) 監査範囲及び監査目的の達成に関する監査報告書のレビュー及び承認

c) 監査所見に対処するためにとった処置の有効性のレビュー
d) 関連する利害関係者への監査報告の配付

e) フォローアップ監査の必要性の決定

必要に応じて,監査プログラムをマネジメントする人は,次の事項を考慮することが望ましい.

— communicating audit results and best practices to other areas of the organization, and
— the implications for other processes.

5.5.7 Managing and maintaining audit programme records

The individual(s) managing the audit programme should ensure that audit records are generated, managed and maintained to demonstrate the implementation of the audit programme. Processes should be established to ensure that any information security and confidentiality needs associated with the audit records are addressed.

Records can include the following:
a) Records related to the audit programme, such as:
   — schedule of audits;
   — audit programme objectives and extent;

   — those addressing audit programme risks and opportunities, and relevant external and internal issues;
   — reviews of the audit programme effective-

— 監査結果及びベストプラクティスを組織の他の領域に伝達すること
— 他のプロセスへの影響

5.5.7　監査プログラムの記録の管理及び維持

　監査プログラムをマネジメントする人は，監査プログラムの実施を実証するために監査記録を作成し，管理し，維持することを確実にすることが望ましい．監査記録に付随する情報セキュリティ及び機密保持に関するいかなるニーズにも対処することを確実にするためのプロセスを確立することが望ましい．

　記録には，次の事項を含み得る．
a)　監査プログラムに関係する，次のような記録

— 監査のスケジュール
— 監査プログラムの目的及び監査プログラムの及ぶ領域
— 監査プログラムのリスク及び機会に対処する事項，並びに関連する外部及び内部の課題

— 監査プログラムの有効性のレビュー

ness.

b) Records related to each audit, such as:
- audit plans and audit reports;
- objective audit evidence and findings;
- nonconformity reports;
- corrections and corrective action reports;
- audit follow-up reports.

c) Records related to the audit team covering topics such as:
- competence and performance evaluation of the audit team members;
- criteria for the selection of audit teams and team members and formation of audit teams;
- maintenance and improvement of competence.

The form and level of detail of the records should demonstrate that the objectives of the audit programme have been achieved.

5.6 Monitoring audit programme

The individual(s) managing the audit programme should ensure the evaluation of:

b) 各監査に関係する,次のような記録
 — 監査計画及び監査報告書
 — 客観的な監査証拠及び監査所見
 — 不適合報告書
 — 修正及び是正処置報告書
 — 監査のフォローアップ報告書
c) 次のような事項を含む,監査チームに関係する記録
 — 監査チームメンバーの力量及びパフォーマンスの評価
 — 監査チーム及び監査チームメンバーの選定並びに監査チームの編成に関する基準

 — 力量の維持及び向上

　記録の形式及び詳細さのレベルは,監査プログラムの目的を達成していることを実証できるものであることが望ましい.

5.6 監査プログラムの監視
　監査プログラムをマネジメントする人は,次の事項の評価を確実にすることが望ましい.

a) whether schedules are being met and audit programme objectives are being achieved;
b) the performance of the audit team members including the audit team leader and the technical experts;
c) the ability of the audit teams to implement the audit plan;
d) feedback from audit clients, auditees, auditors, technical experts and other relevant parties;
e) sufficiency and adequacy of documented information in the whole audit process.

Some factors can indicate the need to modify the audit programme. These can include changes to:

— audit findings;
— demonstrated level of auditee's management system effectiveness and maturity;
— effectiveness of the audit programme;
— audit scope or audit programme scope;
— the auditee's management system;
— standards, and other requirements to which the organization is committed;

a) スケジュールを守り,監査プログラムの目的を達成しているかどうか.
b) 監査チームメンバーのパフォーマンス.これには,監査チームのリーダー及び技術専門家を含む.
c) 監査計画を履行する監査チームの能力

d) 監査依頼者,被監査者,監査員,技術専門家,及びその他関係者からのフィードバック

e) 監査プロセス全体における文書化した情報が十分であること及び妥当であること

　監査プログラムの修正の必要性を示し得る,幾つかの要因がある.これらの要因には,次の事項に対する変更を含み得る.
— 監査所見
— 被監査者のマネジメントシステムの有効性及び成熟度の,実証されたレベル
— 監査プログラムの有効性
— 監査範囲又は監査プログラムの範囲
— 被監査者のマネジメントシステム
— 規格,及び被監査者である組織がコミットメントするその他の要求事項

— external providers;
— identified conflicts of interest;
— the audit client's requirements.

5.7 Reviewing and improving audit programme

The individual(s) managing the audit programme and the audit client should review the audit programme to assess whether its objectives have been achieved. Lessons learned from the audit programme review should be used as inputs for the improvement of the programme.

The individual(s) managing the audit programme should ensure the following:
— review of the overall implementation of the audit programme;
— identification of areas and opportunities for improvement;
— application of changes to the audit programme if necessary;
— review of the continual professional development of auditors, in accordance with **7.6**;
— reporting of the results of the audit pro-

— 外部提供者
— 特定した利害抵触
— 監査依頼者の要求事項

5.7 監査プログラムのレビュー及び改善

　監査プログラムをマネジメントする人及び監査依頼者は，監査プログラムの目的を達成しているかどうかを評価するために，監査プログラムをレビューすることが望ましい．監査プログラムのレビューから得た知見は，プログラムの改善のインプットとして使用することが望ましい．

　監査プログラムをマネジメントする人は，次の事項を確実にすることが望ましい．
— 監査プログラムの全体的な履行のレビュー

— 改善の領域及び改善の機会の特定

— 必要な場合，監査プログラムに対する変更の適用
— **7.6** に従った，監査員の専門能力の継続的開発のレビュー
— 監査プログラムの結果の報告，並びに適宜，監

gramme and review with the audit client and relevant interested parties, as appropriate.

The audit programme review should consider the following:
a) results and trends from audit programme monitoring;
b) conformity with audit programme processes and relevant documented information;
c) evolving needs and expectations of relevant interested parties;
d) audit programme records;
e) alternative or new auditing methods;
f) alternative or new methods to evaluate auditors;
g) effectiveness of the actions to address the risks and opportunities, and internal and external issues associated with the audit programme;
h) confidentiality and information security issues relating to the audit programme.

6 Conducting an audit
6.1 General

査依頼者及び関連する利害関係者とのレビュー

　監査プログラムのレビューでは，次の事項を考慮することが望ましい．

a)　監査プログラムの監視の結果及びその傾向

b)　監査プログラムのプロセス及び関連する文書化した情報との適合

c)　関連する利害関係者から新たに出てきたニーズ及び期待

d)　監査プログラムの記録

e)　代わりの又は新規の監査方法

f)　代わりの又は新規の，監査員を評価する方法

g)　監査プログラムに付随する，リスク及び機会並びに内部及び外部の課題に対処する活動の有効性

h)　監査プログラムに関係する機密保持及び情報セキュリティ上の課題

6　監査の実施
6.1　一般

This clause contains guidance on preparing and conducting a specific audit as part of an audit programme. **Figure 2** provides an overview of the activities performed in a typical audit. The extent to which the provisions of this clause are applicable depends on the objectives and scope of the specific audit.

6.2 Initiating audit
6.2.1 General

The responsibility for conducting the audit should remain with the assigned audit team leader (see **5.5.5**) until the audit is completed (see **6.6**).

To initiate an audit, the steps in **Figure 1** should be considered; however, the sequence can differ depending on the auditee, processes and specific circumstances of the audit.

6.2.2 Establishing contact with auditee

The audit team leader should ensure that contact is made with the auditee to:

a) confirm communication channels with the auditee's representatives;

この箇条では，監査プログラムの一部としての個別の監査の準備及び実施の手引を示す．**図2**は，典型的な監査において実施される活動の概要を示す．この箇条をどの程度適用するかは，個別の監査の目的及び範囲によって異なる．

6.2 監査の開始
6.2.1 一般
監査実施の責任は，監査が完了（**6.6**参照）するまで，割り当てられた監査チームリーダー（**5.5.5**参照）が負うことが望ましい．

監査を開始するために，**図1**のステップを考慮することが望ましい．ただし，被監査者，プロセス及び監査の個別の周辺状況によってステップの順序は異なり得る．

6.2.2 被監査者との連絡の確立
監査チームリーダーは，次の事項のために被監査者との連絡を確実にすることが望ましい．

a) 被監査者の代表者とのコミュニケーションチャネルを確認する．

b) confirm the authority to conduct the audit;

c) provide relevant information on the audit objectives, scope, criteria, methods and audit team composition, including any technical experts;

d) request access to relevant information for planning purposes including information on the risks and opportunities the organization has identified and how they are addressed;

e) determine applicable statutory and regulatory requirements and other requirements relevant to the activities, processes, products and services of the auditee;

f) confirm the agreement with the auditee regarding the extent of the disclosure and the treatment of confidential information;

g) make arrangements for the audit including the schedule;

h) determine any location-specific arrangements for access, health and safety, security, confidentiality or other;

i) agree on the attendance of observers and the need for guides or interpreters for the audit team;

6　監査の実施

b) 監査を行うための権限を確認する．
c) 監査の目的，範囲，基準，方法及び技術専門家を含む監査チームの構成に関連する情報を提供する．

d) 計画策定の目的のために関連情報へのアクセスを要請する．関連情報には，組織が特定したリスク及び機会，並びにそれらへどのように対処するかに関する情報を含む．
e) 適用される法令・規制要求事項，並びに被監査者の活動，プロセス，製品及びサービスに関連するその他の要求事項を決定する．

f) 情報公開の範囲及び機密情報の取扱いに関して，被監査者との合意を確認する．

g) スケジュールを含め，監査のための手配をする．
h) それぞれの場所に固有の手配事項として，アクセス，安全衛生，セキュリティ，機密保持，その他について決定する．
i) オブザーバの参加，及び監査チームのための案内役又は通訳者の必要性について合意する．

j) determine any areas of interest, concern or risks to the auditee in relation to the specific audit;

k) resolve issues regarding composition of the audit team with the auditee or audit client.

6.2.3 Determining feasibility of audit

The feasibility of the audit should be determined to provide reasonable confidence that the audit objectives can be achieved.

The determination of feasibility should take into consideration factors such as the availability of the following:

a) sufficient and appropriate information for planning and conducting the audit;

b) adequate cooperation from the auditee;

c) adequate time and resources for conducting the audit.

> NOTE Resources include access to adequate and appropriate information and communication technology.

j) 個別の監査に関係して，被監査者に対する利害，懸念事項，又はリスクの，あらゆる領域を決定する．

k) 被監査者又は監査依頼者とともに，監査チームの構成に関する課題を解決する．

6.2.3 監査の実施可能性の決定

監査目的を達成し得るという合理的な確信を得るために，監査の実施可能性を決定することが望ましい．

実施可能性の決定には，次の要素が利用可能であるかどうかを考慮に入れることが望ましい．

a) 監査の計画を策定し，監査を行うための十分かつ適切な情報

b) 被監査者の十分な協力

c) 監査を行うための十分な時間及び資源

> **注記** 資源には，十分かつ適切な情報通信技術へのアクセスを含む．

Where the audit is not feasible, an alternative should be proposed to the audit client, in agreement with the auditee.

6.3 Preparing audit activities

6.3.1 Performing review of documented information

The relevant management system documented information of the auditee should be reviewed in order to:

— gather information to understand the auditee's operations and to prepare audit activities and applicable audit work documents (see **6.3.4**), e.g. on processes, functions;
— establish an overview of the extent of the documented information to determine possible conformity to the audit criteria and detect possible areas of concern, such as deficiencies, omissions or conflicts.

The documented information should include, but not be limited to: management system documents and records, as well as previous audit reports. The review should take into account the context of the

監査が実施不可能な場合,被監査者との合意の上で,監査依頼者に代替案を提示することが望ましい.

6.3 監査活動の準備
6.3.1 文書化した情報のレビューの実施

次の事項のために,関連する被監査者のマネジメントシステムの文書化した情報をレビューすることが望ましい.
— 被監査者の運用を理解し,監査活動の準備をするための情報,及び適用される監査作業文書(**6.3.4** 参照),例えばプロセス,機能などに関する監査作業文書を集める.
— 監査基準への適合の可能性を決定し,不備,脱落,不一致などのような潜在的な懸念領域を検出するために,文書化した情報の範囲の全体像を確立する.

文書化した情報には,マネジメントシステム文書及び記録,並びに前回までの監査報告を含めることが望ましいが,これらに限定されない.レビューでは,被監査者の組織の規模,性質,複雑さ,並びに

auditee's organization, including its size, nature and complexity, and its related risks and opportunities. It should also take into account the audit scope, criteria and objectives.

NOTE   Guidance on how to verify information is provided in **A.5**.

6.3.2   Audit planning
6.3.2.1   Risk-based approach to planning
The audit team leader should adopt a risk-based approach to planning the audit based on the information in the audit programme and the documented information provided by the auditee.

Audit planning should consider the risks of the audit activities on the auditee's processes and provide the basis for the agreement among the audit client, audit team and the auditee regarding the conduct of the audit. Planning should facilitate the efficient scheduling and coordination of the audit activities in order to achieve the objectives effectively.

関連するリスク及び機会を含む，組織の状況を考慮に入れることが望ましい．また，監査範囲，監査基準及び監査目的も考慮に入れることが望ましい．

> 注記　どのように情報を検証するかについての手引を **A.5** に示す．

6.3.2　監査計画の策定
6.3.2.1　計画策定へのリスクに基づくアプローチ

　監査チームリーダーは，監査プログラム中の情報及び被監査者から提供される文書化した情報に基づいて，監査計画の策定にリスクに基づくアプローチを採用することが望ましい．

　監査計画の策定は，被監査者のプロセスに関する監査活動のリスクを考慮することが望ましく，また，監査依頼者，監査チーム及び被監査者の間で，監査の実施に関する合意形成の基礎を提示することが望ましい．監査計画の策定によって，監査目的を有効に達成するための監査活動の効率的なスケジュールの策定及び調整を行いやすくすることが望ましい．

The amount of detail provided in the audit plan should reflect the scope and complexity of the audit, as well as the risk of not achieving the audit objectives. In planning the audit, the audit team leader should consider the following:

a) the composition of the audit team and its overall competence;

b) the appropriate sampling techniques (see **A.6**);

c) opportunities to improve the effectiveness and efficiency of the audit activities;

d) the risks to achieving the audit objectives created by ineffective audit planning;

e) the risks to the auditee created by performing the audit.

Risks to the auditee can result from the presence of the audit team members adversely influencing the auditee's arrangements for health and safety, environment and quality, and its products, services, personnel or infrastructure (e.g. contamination in clean room facilities).

For combined audits, particular attention should be given to the interactions between operational

監査計画に提示する詳細さの程度は，監査の範囲及び複雑さ，並びに監査目的を達成できないリスクを反映していることが望ましい．監査計画の策定に当たって，監査チームリーダーは次の事項を考慮することが望ましい．

a) 監査チームの構成及びその全体としての力量

b) 適切なサンプリング技法（**A.6** 参照）

c) 監査活動の有効性及び効率を改善する機会

d) 有効でない監査計画の策定によって生み出される，監査目的の達成に対するリスク

e) 監査の実施によって生み出される，被監査者に対するリスク

　被監査者に対するリスクとなり得ることとして，監査チームメンバーの存在が，被監査者の安全衛生，環境及び品質に悪影響を与えること，並びにその製品，サービス，要員又はインフラストラクチャに対して脅威となることがある（例えば，クリーンルーム設備内の汚染）．

　複合監査については，異なるマネジメントシステムの運用プロセス間の相互関係，並びにあらゆる競

processes and any competing objectives and priorities of the different management systems.

6.3.2.2 Audit planning details

The scale and content of the audit planning can differ, for example, between initial and subsequent audits, as well as between internal and external audits. Audit planning should be sufficiently flexible to permit changes which can become necessary as the audit activities progress.

Audit planning should address or reference the following:

a) the audit objectives;
b) the audit scope, including identification of the organization and its functions, as well as processes to be audited;
c) the audit criteria and any reference documented information;
d) the locations (physical and virtual), dates, expected time and duration of audit activities to be conducted, including meetings with the auditee's management;
e) the need for the audit team to familiarize

合する目的及びそれらの優先順位に対して特別の注意を払うことが望ましい．

6.3.2.2　監査計画の策定の詳細

　監査計画の策定の規模及び内容は，例えば初回の監査とその後に続く監査とで異なり得る．また，内部監査と外部監査とでも同様である．監査計画の策定は，監査活動の進行に伴って必要となり得る変更を許容する十分な柔軟性をもっていることが望ましい．

　監査計画の策定は，次の事項に対処するか，又はその参照先を示すことが望ましい．
a)　監査目的
b)　監査範囲．これには，監査の対象となる組織及び組織の機能並びにプロセスの特定を含む．

c)　監査基準及びあらゆる参照となる文書化した情報
d)　監査活動を行う場所（物理的及び仮想的），日程，予定時間及び予定の工数．これには，被監査者の管理層との会議を含む．

e)　監査チームが，被監査者の施設及びプロセスを

themselves with auditee's facilities and processes (e.g. by conducting a tour of physical location(s), or reviewing information and communication technology);

f) the audit methods to be used, including the extent to which audit sampling is needed to obtain sufficient audit evidence;

g) the roles and responsibilities of the audit team members, as well as guides and observers or interpreters;

h) the allocation of appropriate resources based upon consideration of the risks and opportunities related to the activities that are to be audited.

Audit planning should take into account, as appropriate:

— identification of the auditee's representative(s) for the audit;
— the working and reporting language of the audit where this is different from the language of the auditor or the auditee or both;
— the audit report topics;
— logistics and communications arrangements,

理解する必要性(例えば,物理的な場所の視察,又は情報通信技術のレビューによって)

f) 使用する監査方法.これには,十分な監査証拠を得るために必要な監査サンプリングの程度を含む.
g) 監査チームメンバーの役割及び責任.案内役,及びオブザーバ又は通訳者の役割及び責任も同様である.
h) 監査対象となる活動に関係したリスク及び機会の考慮に基づいた,適切な資源の配分

　監査計画の策定には,必要に応じて,次の事項を考慮に入れることが望ましい.
— 監査に対する被監査者の代表者の特定

— 監査の作業及び報告に用いる言語で,監査員若しくは被監査者又はその両方の言語と異なる場合
— 監査報告書の記載項目
— 監査の後方支援(logistics)及びコミュニケー

including specific arrangements for the locations to be audited;
— any specific actions to be taken to address risks to achieving the audit objectives and opportunities arising;
— matters related to confidentiality and information security;
— any follow-up actions from a previous audit or other source(s) e.g. lessons learned, project reviews;
— any follow-up activities to the planned audit;

— coordination with other audit activities, in case of a joint audit.

Audit plans should be presented to the auditee. Any issues with the audit plans should be resolved between the audit team leader, the auditee and, if necessary, the individual(s) managing the audit programme.

6.3.3 Assigning work to audit team

The audit team leader, in consultation with the audit team, should assign to each team member re-

ションに関する手配事項.これには,監査の対象となる場所に対する個別の手配を含む.
— 監査目的の達成に対するリスク及び発生する機会に対処してとるあらゆる個別の処置

— 機密保持及び情報セキュリティに関係する事項

— 前回の監査又はその他の情報源,例えば,得られた知見,プロジェクトレビューなどに対するあらゆるフォローアップ処置
— 計画した監査に対するあらゆるフォローアップ活動
— 合同監査の場合,他の監査活動との調整

監査計画は,被監査者に提示することが望ましい.監査計画についてのあらゆる課題は,監査チームリーダー,被監査者,及び必要があれば監査プログラムをマネジメントする人との間で解決することが望ましい.

6.3.3 監査チームへの作業の割当て

監査チームリーダーは,監査チームと協議し,チームメンバー各々に,個別のプロセス,活動,機能

sponsibility for auditing specific processes, activities, functions or locations and, as appropriate, authority for decision-making. Such assignments should take into account the impartiality and objectivity and competence of auditors and the effective use of resources, as well as different roles and responsibilities of auditors, auditors-in-training and technical experts.

Audit team meetings should be held, as appropriate, by the audit team leader in order to allocate work assignments and decide possible changes. Changes to the work assignments can be made as the audit progresses in order to ensure the achievement of the audit objectives.

6.3.4 Preparing documented information for audit

The audit team members should collect and review the information relevant to their audit assignments and prepare documented information for the audit, using any appropriate media. The documented information for the audit can include but is not limited to:

又は場所を監査する責任を，及び該当する場合は，意思決定の権限を割り当てることが望ましい．このような割当てを行う際には，監査員の公平性及び客観性並びに力量を考慮に入れるとともに，資源の有効な利用並びに監査員，訓練中の監査員及び技術専門家それぞれの異なる役割及び責任を考慮に入れることが望ましい．

監査チーム会議は，作業分担の割当て及び起こり得る変更について決定するために，適宜，監査チームリーダーが開催することが望ましい．監査目的の達成を確実にするために，監査の進行に伴い，作業分担を変更することができる．

6.3.4 監査のための文書化した情報の作成

監査チームメンバーは，監査の割当てに関連する情報を収集及びレビューし，並びに適切な媒体を用いて，その監査のための文書化した情報を作成することが望ましい．監査のための文書化した情報には，次の事項を含み得るが，これらに限らない．

a) physical or digital checklists;

b) audit sampling details;
c) audio visual information.

The use of these media should not restrict the extent of audit activities, which can change as a result of information collected during the audit.

NOTE Guidance on preparing audit work documents is given in **A.13**.

Documented information prepared for, and resulting from, the audit should be retained at least until audit completion, or as specified in the audit programme. Retention of documented information after audit completion is described in **6.6**. Documented information created during the audit process involving confidential or proprietary information should be suitably safeguarded at all times by the audit team members.

6.4 Conducting audit activities

a) チェックリスト．これには，物理的又は電子的なものがある．
b) 監査サンプリングの詳細
c) 視聴覚情報

　これらの媒体の利用が，監査活動の及ぶ領域を限定しないことが望ましい．この監査活動の及ぶ領域は，監査中に収集した情報の結果として変化し得る．

　　注記　監査作業文書の作成に関する手引を**A.13**に示す．

　監査のため又は監査の結果として作成した文書化した情報は，少なくとも監査が完了するまで，又は監査プログラムで定めたとおりに，保持することが望ましい．監査完了後の文書化した情報の保持は，**6.6**に示す．監査プロセス中に作成した，機密情報又は所有者情報を含む文書化した情報は，監査チームメンバーが常に適切な安全対策を施すことが望ましい．

6.4　監査活動の実施

6.4.1 General

Audit activities are normally conducted in a defined sequence as indicated in **Figure 1**. This sequence may be varied to suit the circumstances of specific audits.

6.4.2 Assigning roles and responsibilities of guides and observers

Guides and observers may accompany the audit team with approvals from the audit team leader, audit client and/or auditee, if required. They should not influence or interfere with the conduct of the audit. If this cannot be assured, the audit team leader should have the right to deny observers from being present during certain audit activities.

For observers, any arrangements for access, health and safety, environmental, security and confidentiality should be managed between the audit client and the auditee.

Guides, appointed by the auditee, should assist the audit team and act on the request of the audit

6.4.1 一般

監査活動は，通常，図1で示す，定めた順序で行う．この順序は，個別の監査の状況に合わせて変えてよい．

6.4.2 案内役及びオブザーバの役割及び責任の割当て

案内役及びオブザーバは，必要があれば，監査チームリーダー，監査依頼者及び／又は被監査者の承認を得て，監査チームに同行してよい．案内役及びオブザーバは，監査の実施に影響を及ぼしたり，妨害をしたりしないことが望ましい．これが保証できない場合，監査チームリーダーは，オブザーバの一定の監査活動への参加を拒否する権利をもつことが望ましい．

オブザーバについては，アクセス，安全衛生，環境，セキュリティ及び機密保持に関するあらゆる取決めを，監査依頼者と被監査者との間でマネジメントすることが望ましい．

被監査者に指名された案内役は，監査チームを手助けし，監査チームリーダー又は担当する監査員の

team leader or the auditor to which they have been assigned. Their responsibilities should include the following:

a) assisting the auditors in identifying individuals to participate in interviews and confirming timings and locations;

b) arranging access to specific locations of the auditee;

c) ensuring that rules concerning location-specific arrangements for access, health and safety, environmental, security, confidentiality and other issues are known and respected by the audit team members and observers and any risks are addressed;

d) witnessing the audit on behalf of the auditee, when appropriate;

e) providing clarification or assisting in collecting information, when needed.

6.4.3 Conducting opening meeting

The purpose of the opening meeting is to:

a) confirm the agreement of all participants (e.g. auditee, audit team) to the audit plan;

b) introduce the audit team and their roles;

要請に応じて行動することが望ましい．案内役の責任には，次の事項を含めることが望ましい．

a) インタビューに参加する個人の特定並びにインタビューのタイミング及び場所の確認において監査員を手助けする．
b) 被監査者の特定の場所へのアクセスを手配する．
c) アクセスに関する場所固有の取決め，安全衛生，環境，セキュリティ，機密保持，及びその他の課題に関わる規則について，監査チームメンバー及びオブザーバへの周知及び順守，並びにあらゆるリスクへの対処を確実にする．

d) 適宜，被監査者の代理として監査に立ち会う．

e) 必要があれば，情報収集において不明な点を明らかにし，又は情報収集の手助けをする．

6.4.3 初回会議の実施

初回会議の目的は，次の事項を行うことである．
a) 監査計画に対して，全ての参加者（例えば，被監査者，監査チーム）の合意を確認する．
b) 監査チーム及びその役割を紹介する．

c)   ensure that all planned audit activities can be performed.

An opening meeting should be held with the auditee's management and, where appropriate, those responsible for the functions or processes to be audited. During the meeting, an opportunity to ask questions should be provided.

The degree of detail should be consistent with the familiarity of the auditee with the audit process. In many instances, e.g. internal audits in a small organization, the opening meeting may simply consist of communicating that an audit is being conducted and explaining the nature of the audit.

For other audit situations, the meeting may be formal and records of attendance should be retained. The meeting should be chaired by the audit team leader.

Introduction of the following should be considered, as appropriate:

—   other participants, including observers and

c) 全ての計画した監査活動を行い得ることを確実にする.

　初回会議は,被監査者の管理層,及び適切な場合には,監査の対象となる機能又はプロセスの責任者が,参加して開催することが望ましい.会議中,質問をする機会を与えることが望ましい.

　詳細さの程度は,被監査者の監査プロセスへの精通度に合致したものであることが望ましい.多くの場合には,例えば,小規模な組織での内部監査では,初回会議は,単に監査がこれから行われることを伝え,その監査の性質を説明するだけでもよい.

　それ以外の監査の場合では,初回会議は正式なものとしてよい.その場合には,出席者の記録を保持することが望ましい.初回会議では,監査チームリーダーが議長を務めることが望ましい.

　次の事項の紹介を適宜考慮することが望ましい.

― オブザーバ及び案内役,通訳者を含むその他の

guides, interpreters and an outline of their roles;
— the audit methods to manage risks to the organization which may result from the presence of the audit team members.

Confirmation of the following items should be considered, as appropriate:
— the audit objectives, scope and criteria;
— the audit plan and other relevant arrangements with the auditee, such as the date and time for the closing meeting, any interim meetings between the audit team and the auditee's management, and any change(s) needed;
— formal communication channels between the audit team and the auditee;
— the language to be used during the audit;
— the auditee being kept informed of audit progress during the audit;
— the availability of the resources and facilities needed by the audit team;
— matters relating to confidentiality and information security;

参加者，並びにそれぞれの役割の概要

— 組織に対するリスクをマネジメントする監査方法．このリスクは，監査チームメンバーの存在に起因するかもしれない．

次の事項の確認を適宜考慮することが望ましい．

— 監査の目的，範囲及び基準
— 監査計画及び他の関連する被監査者との取決め，並びに必要な変更．取決めとは，例えば，監査チームと被監査者の管理層との間の，最終会議及び中間会議の日時．

— 監査チームと被監査者との間の正式なコミュニケーションチャネル
— 監査に使用する言語
— 監査中は，監査の進捗状況を被監査者に常に知らせること
— 監査チームが必要とする資源及び施設が利用可能であること
— 機密保持及び情報セキュリティに関係する事項

— relevant access, health and safety, security, emergency and other arrangements for the audit team;
— activities on site that can impact the conduct of the audit.

The presentation of information on the following items should be considered, as appropriate:
— the method of reporting audit findings including criteria for grading, if any;
— conditions under which the audit may be terminated;
— how to deal with possible findings during the audit;
— any system for feedback from the auditee on the findings or conclusions of the audit, including complaints or appeals.

6.4.4 Communicating during audit

During the audit, it may be necessary to make formal arrangements for communication within the audit team, as well as with the auditee, the audit client and potentially with external interested parties (e.g. regulators), especially where statuto-

— 監査チームに対する，関連するアクセス，安全衛生，セキュリティ，緊急時及びその他の取決め
— 監査の実施に影響し得る現地（サイト）での活動

次の事項に関する情報の提示を適宜考慮することが望ましい．
— 存在する場合，格付基準を含む，監査所見の報告の方法
— 監査を打ち切ってよい条件

— 監査中に出てくる可能性のある所見の取扱い方

— 苦情又は異議申立てを含む監査所見又は監査結論についての，被監査者からのフィードバックのためのシステム

6.4.4　監査中のコミュニケーション

　監査中，監査チーム内，並びに被監査者，監査依頼者及び必要であれば外部の利害関係者（例えば，規制当局）とのコミュニケーションについて，正式な取決めが必要となることがある．特に，法令・規制要求事項の不適合について，報告が義務として求

ry and regulatory requirements require mandatory reporting of nonconformities.

The audit team should confer periodically to exchange information, assess audit progress and reassign work between the audit team members, as needed.

During the audit, the audit team leader should periodically communicate the progress, any significant findings and any concerns to the auditee and audit client, as appropriate. Evidence collected during the audit that suggests an immediate and significant risk should be reported without delay to the auditee and, as appropriate, to the audit client. Any concern about an issue outside the audit scope should be noted and reported to the audit team leader, for possible communication to the audit client and auditee.

Where the available audit evidence indicates that the audit objectives are unattainable, the audit team leader should report the reasons to the audit client and the auditee to determine appropriate ac-

監査チームは，情報交換，監査進捗状況の評価，及び必要な場合には，監査チームメンバー間での作業の再割当てのために，定期的に打ち合せることが望ましい．

　監査中，監査チームリーダーは，進捗状況，あらゆる重大な所見及びあらゆる懸念事項を，被監査者及び適宜，監査依頼者に，定期的に連絡することが望ましい．監査中に収集した証拠で緊急かつ重大なリスクを示唆するものがあれば，被監査者及び適宜，監査依頼者に，遅滞なく報告することが望ましい．監査範囲外の課題に関するいかなる懸念も，監査依頼者及び被監査者に連絡をとる場合に備えて，メモをとり，監査チームリーダーに報告することが望ましい．

　入手できる監査証拠から監査目的が達成できないことが明確になった場合には，監査チームリーダーは，適切な処置を決定するために，監査依頼者及び被監査者へ監査目的が達成できない理由を報告する

tion. Such action may include changes to audit planning, the audit objectives or audit scope, or termination of the audit.

Any need for changes to the audit plan which may become apparent as auditing activities progress should be reviewed and accepted, as appropriate, by both the individual(s) managing the audit programme and the audit client, and presented to the auditee.

6.4.5 Audit information availability and access

The audit methods chosen for an audit depend on the defined audit objectives, scope and criteria, as well as duration and location. The location is where the information needed for the specific audit activity is available to the audit team. This may include physical and virtual locations.

Where, when and how to access audit information is crucial to the audit. This is independent of where the information is created, used and/or stored. Based on these issues, the audit methods need to

ことが望ましい．このような処置には，監査計画の変更，監査目的若しくは監査範囲の変更，又は監査の打切りを含めてもよい．

監査活動の進捗に伴って監査計画の変更の必要が明らかになった場合には，このような変更の必要性を，監査プログラムをマネジメントする人及び監査依頼者の双方が適宜レビュー及び受諾し，被監査者に報告することが望ましい．

6.4.5 監査情報の入手可能性及びアクセス

監査のために選択する監査方法は，定められた監査の目的，範囲及び基準，並びに期間及び場所による．場所とは，特定の監査活動に必要な情報を監査チームが入手することができる所である．これには，物理的及び仮想的な場所を含めてもよい．

どこで，いつ，どのように監査情報にアクセスできるかという点は監査において極めて重要である．これは，情報が生成，利用及び／又は保管される場所に影響を受けない．これらの課題に基づいて監

be determined (see **Table A.1**). The audit can use a mixture of methods. Also, audit circumstances may mean that the methods need to change during the audit.

6.4.6 Reviewing documented information while conducting audit

The auditee's relevant documented information should be reviewed to:
— determine the conformity of the system, as far as documented, with audit criteria;
— gather information to support the audit activities.

NOTE   Guidance on how to verify information is provided in **A.5**.

The review may be combined with the other audit activities and may continue throughout the audit, providing this is not detrimental to the effectiveness of the conduct of the audit.

If adequate documented information cannot be provided within the time frame given in the audit

査方法を決定する必要がある（**表 A.1** 参照）．監査は，複数の方法を組み合わせて使用することができる．また，監査をめぐる状況から，その方法を監査中に変更する意味合いが生じる場合がある．

6.4.6 監査の実施中の，文書化した情報のレビュー

被監査者の，関連する文書化した情報は，次の事項を行うために，レビューすることが望ましい．
— 文書化された範囲で，監査基準に対する，システムの適合性を決定する．
— 監査活動を支援する情報を集める．

> **注記** どのように情報を検証するかについての手引を **A.5** に示す．

レビューは，その他の監査活動と組み合わせてよく，また，監査の実施の有効性に支障を来さなければ，監査を通じて継続して行ってよい．

監査計画で与えられた時間枠内に，十分な文書化した情報が提供されなかった場合には，監査チーム

plan, the audit team leader should inform both the individual(s) managing the audit programme and the auditee. Depending on the audit objectives and scope, a decision should be made as to whether the audit should be continued or suspended until documented information concerns are resolved.

6.4.7 Collecting and verifying information

During the audit, information relevant to the audit objectives, scope and criteria, including information relating to interfaces between functions, activities and processes should be collected by means of appropriate sampling and should be verified, as far as practicable.

NOTE 1 For verifying information see **A.5**.

NOTE 2 Guidance on sampling is given in **A.6**.

Only information that can be subject to some degree of verification should be accepted as audit evidence. Where the degree of verification is low the auditor should use their professional judgement to determine the degree of reliance that can be placed

リーダーは，監査プログラムをマネジメントする人及び被監査者の双方に，その旨を知らせることが望ましい．監査の目的及び範囲によって，監査を続行するか，又は文書化した情報に関する懸念が解決するまで中断するか，について決定することが望ましい．

6.4.7　情報の収集及び検証

監査中は，監査の目的，範囲及び基準に関連する情報を，機能，活動及びプロセス間のインタフェースに関係する情報を含めて，実践できる限り，適切なサンプリング手段によって収集し，検証することが望ましい．

> 注記1　情報の検証については A.5 を参照．
> 注記2　サンプリングについての手引を A.6 に示す．

ある程度の検証の対象となり得る情報だけを監査証拠として採用することが望ましい．検証の程度が低い場合には，その証拠にどの程度の信頼を置き得るかを決定するために，監査員は各自の専門的な判断を用いることが望ましい．監査所見を導く監査証

on it as evidence. Audit evidence leading to audit findings should be recorded. If, during the collection of objective evidence, the audit team becomes aware of any new or changed circumstances, or risks or opportunities, these should be addressed by the team accordingly.

Figure 2 provides an overview of a typical process, from collecting information to reaching audit conclusions.

拠は，記録することが望ましい．客観的証拠の収集中に監査チームが，何らかの新しい若しくは変化した状況，又はリスク若しくは機会に気付いたならば，監査チームはしかるべくこれらに対処することが望ましい．

　情報収集から監査結論に至るまでの典型的なプロセスの概要を図 2 に示す．

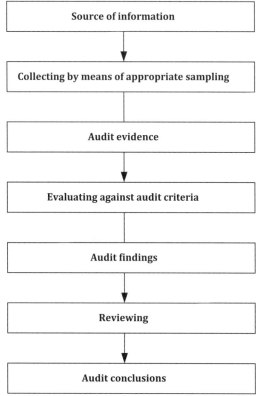

Figure 2 — Overview of a typical process of collecting and verifying information

Methods of collecting information include, but are not limited to the following:

— interviews;
— observations;

6 監査の実施　　　175

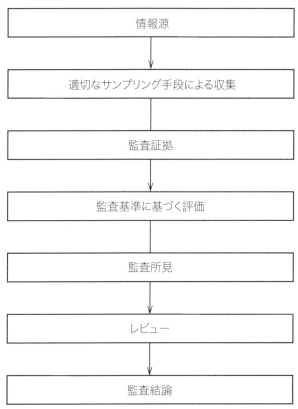

図2 − 情報の収集及び検証の典型的なプロセスの概要

　情報を収集する方法には，次の事項を含むが，これらに限定されない．
— インタビュー
— 観察

— review of documented information.

NOTE 3   Guidance on selecting sources of information and observation is given in **A.14**.

NOTE 4   Guidance on visiting the auditee's location is given in **A.15**.

NOTE 5   Guidance on conducting interviews is given in **A.17**.

6.4.8   Generating audit findings

Audit evidence should be evaluated against the audit criteria in order to determine audit findings. Audit findings can indicate conformity or nonconformity with audit criteria. When specified by the audit plan, individual audit findings should include conformity and good practices along with their supporting evidence, opportunities for improvement, and any recommendations to the auditee.

Nonconformities and their supporting audit evidence should be recorded.

Nonconformities can be graded depending on the

― 文書化した情報のレビュー

> 注記3　情報源の選択,及び観察についての手引を **A.14** に示す.
> 注記4　被監査者の場所を訪問する際の手引を **A.15** に示す.
> 注記5　インタビュー実施についての手引を **A.17** に示す.

6.4.8　監査所見の作成

　監査所見を決定するために,監査基準に照らして監査証拠を評価することが望ましい.監査所見では,監査基準に対して適合又は不適合のいずれかを示すことができる.監査計画で規定されている場合には,個々の監査所見には,根拠となる証拠を伴った適合性及び優れた実践事例,改善の機会,並びに被監査者に対するあらゆる提言を含めることが望ましい.

　不適合及びその根拠となる監査証拠は,記録しておくことが望ましい.

　不適合は,組織の状況及びそのリスクによって格

context of the organization and its risks. This grading can be quantitative (e.g. 1 to 5) and qualitative (e.g. minor, major). They should be reviewed with the auditee in order to obtain acknowledgement that the audit evidence is accurate and that the nonconformities are understood. Every attempt should be made to resolve any diverging opinions concerning the audit evidence or findings. Unresolved issues should be recorded in the audit report.

The audit team should meet as needed to review the audit findings at appropriate stages during the audit.

NOTE 1   Additional guidance on the identification and evaluation of audit findings is given in **A.18**.

NOTE 2   Conformity or nonconformity with audit criteria related to statutory or regulatory requirements or other requirements, is sometimes referred to as compliance or non-compliance.

6.4.9 Determining audit conclusions

付けすることが可能である．この格付けは，定量的なもの（例えば，1から5）も定性的なもの（例えば，軽微，重大）もあり得る．不適合は，被監査者とレビューすることが望ましい．これは，監査証拠が正確であること，及び不適合の内容が理解されたことについて被監査者の確認を得るためである．監査証拠又は監査所見に関して意見の相違がある場合には，それを解決するためのあらゆる努力を試みることが望ましい．解決できなかった課題は，監査報告書に記録しておくことが望ましい．

監査中の適切な段階で監査所見をレビューするために，監査チームは，必要に応じて打合せをすることが望ましい．

注記1　監査所見の特定及び評価についての追加の手引を **A.18** に示す．

注記2　法令・規制要求事項又はその他の要求事項に関係する監査基準に対する適合又は不適合は，順守又は不順守と呼ばれることもある．

6.4.9　監査結論の決定

6.4.9.1 Preparation for closing meeting

The audit team should confer prior to the closing meeting in order to:

a) review the audit findings and any other appropriate information collected during the audit, against the audit objectives;

b) agree on the audit conclusions, taking into account the uncertainty inherent in the audit process;

c) prepare recommendations, if specified by the audit plan;

d) discuss audit follow-up, as applicable.

6.4.9.2 Content of audit conclusions

Audit conclusions should address issues such as the following:

a) the extent of conformity with the audit criteria and robustness of the management system, including the effectiveness of the management system in meeting the intended outcomes, the identification of risks and effectiveness of actions taken by the auditee to address risks;

6.4.9.1　最終会議の準備

監査チームは,最終会議に先立って,次の事項を行うために打ち合せることが望ましい.

a) 監査所見及び監査中に収集したその他の適切な情報を,監査目的に照らしてレビューする.

b) 監査プロセスに内在する不確かさを考慮に入れた上で,監査結論について合意する.

c) 監査計画で規定している場合には,提言を作成する.

d) 該当する場合には,監査のフォローアップについて協議する.

6.4.9.2　監査結論の内容

監査結論では,次のような課題に対処することが望ましい.

a) マネジメントシステムの監査基準への適合の程度及びマネジメントシステムの堅ろう(牢)さ.これには,意図した成果を満たすことにおけるマネジメントシステムの有効性,リスクの特定,及び被監査者がリスクに対処するためにとった処置の有効性を含む.

b) the effective implementation, maintenance and improvement of the management system;
c) achievement of audit objectives, coverage of audit scope and fulfilment of audit criteria;
d) similar findings made in different areas that were audited or from a joint or previous audit for the purpose of identifying trends.

If specified by the audit plan, audit conclusions can lead to recommendations for improvement, or future auditing activities.

6.4.10 Conducting closing meeting

A closing meeting should be held to present the audit findings and conclusions.

The closing meeting should be chaired by the audit team leader and attended by the management of the auditee and include, as applicable:
— those responsible for the functions or processes which have been audited;
— the audit client;
— other members of the audit team;

b) マネジメントシステムの有効な，実施，維持及び改善
c) 監査目的の達成，監査範囲の網羅及び監査基準を満たすこと
d) 傾向を特定する目的に役立つ，類似の所見．これらには，異なる領域における監査から得られたもの，又は合同監査若しくは前回までの監査から得られたものがある．

　監査計画に規定している場合には，監査結論を，改善のための提言又は今後の監査活動につなげることができる．

6.4.10　最終会議の実施
　最終会議は，監査所見及び監査結論を提示するために開催することが望ましい．

　最終会議は，監査チームリーダーが議長を務め，被監査者の管理層が出席し，さらに，該当する場合，次の者を含むことが望ましい．
— 監査を受けた機能又はプロセスの責任者

— 監査依頼者
— 監査チームのリーダー以外のメンバー

— other relevant interested parties as determined by the audit client and/or auditee.

If applicable, the audit team leader should advise the auditee of situations encountered during the audit that may decrease the confidence that can be placed in the audit conclusions. If defined in the management system or by agreement with the audit client, the participants should agree on the time frame for an action plan to address audit findings.

The degree of detail should take into account the effectiveness of the management system in achieving the auditee's objectives, including consideration of its context and risks and opportunities.

The familiarity of the auditee with the audit process should also be taken into consideration during the closing meeting, to ensure the correct level of detail is provided to participants.

For some audit situations, the meeting can be formal and minutes, including records of attendance,

― 監査依頼者及び／又は被監査者が決定する，その他の関連する利害関係者

該当する場合，監査チームリーダーは，監査結論に付与し得る信頼性を低下させるかもしれない，監査中に遭遇した状況について，被監査者に知らせることが望ましい．マネジメントシステムに定められているか，又は監査依頼者との合意がある場合，参加者は，監査所見に対処するための処置の計画の期限について合意することが望ましい．

最終会議の詳細さの程度は，被監査者の目的（又は目標）を達成するためにマネジメントシステムの有効性を考慮に入れることが望ましい．これには，被監査者の状況並びにリスク及び機会の考慮を含む．

被監査者の監査プロセスに関する精通度もまた，最終会議において考慮に入れることが望ましい．これは，最終会議を適正なレベルの詳細さで参加者へ提供することを確実にするためである．

監査の位置づけによっては，最終会議が正式なものとなり得る場合がある．その場合には，出席者の

should be kept. In other instances, e.g. internal audits, the closing meeting can be less formal and consist solely of communicating the audit findings and audit conclusions.

As appropriate, the following should be explained to the auditee in the closing meeting:

a) advising that the audit evidence collected was based on a sample of the information available and is not necessarily fully representative of the overall effectiveness of the auditee's processes;

b) the method of reporting;

c) how the audit finding should be addressed based on the agreed process;

d) possible consequences of not adequately addressing the audit findings;

e) presentation of the audit findings and conclusions in such a manner that they are understood and acknowledged by the auditee's management;

f) any related post-audit activities (e.g. implementation and review of corrective actions, addressing audit complaints, appeal process).

記録を含めて議事録を残すことが望ましい．正式なものとしない場合には，例えば内部監査では，最終会議は，より非公式で，単に監査所見及び監査結論を伝えるだけのものになり得る．

　該当する場合には，最終会議では，次の事項を被監査者に説明することが望ましい．

a) 収集した監査証拠は入手可能な情報のサンプルに基づいたものであり，必ずしも，被監査者のプロセスの全体的有効性を完全に表すものではないことを伝える．

b) 報告の方法

c) 合意したプロセスに基づいて，監査所見にどのように対処するのが望ましいか

d) 監査所見に適切に対処しなかった場合に起こり得る結果

e) 監査所見及び監査結論の提示．被監査者の管理層が理解し，認知する方法で行う．

f) 関係する監査後のあらゆる活動（例えば，是正処置の実施及びレビュー，監査に関する苦情への対処，異議申立てのプロセス）

Any diverging opinions regarding the audit findings or conclusions between the audit team and the auditee should be discussed and, if possible, resolved. If not resolved, this should be recorded.

If specified by the audit objectives, opportunities for improvement recommendations may be presented. It should be emphasized that recommendations are not binding.

6.5 Preparing and distributing audit report
6.5.1 Preparing audit report

The audit team leader should report the audit conclusions in accordance with the audit programme. The audit report should provide a complete, accurate, concise and clear record of the audit, and should include or refer to the following:

a) audit objectives;
b) audit scope, particularly identification of the organization (the auditee) and the functions or processes audited;
c) identification of the audit client;
d) identification of audit team and auditee's par-

監査所見又は監査結論に関して，監査チームと被監査者との間に意見の相違があれば，協議し，可能であれば，それを解決することが望ましい．解決できなかったならば，これを記録に残すことが望ましい．

監査目的で規定している場合は，改善の機会についての提言をしてもよい．提言には，拘束力がないことを強調しておくことが望ましい．

6.5 監査報告書の作成及び配付
6.5.1 監査報告書の作成
　監査チームリーダーは，監査プログラムに従って監査結論を報告することが望ましい．監査報告書は，完全で，正確，簡潔かつ明確な監査の記録を提供することが望ましく，次に示す事項を含むか，又はその事項の参照先を示すことが望ましい．

a) 　監査目的
b) 　監査範囲，特に，監査を受けた組織（被監査者）及びその機能又はプロセスの特定
c) 　監査依頼者の特定
d) 　監査チーム及び被監査者の監査参加者の特定

ticipants in the audit;

e) dates and locations where the audit activities were conducted;
f) audit criteria;
g) audit findings and related evidence;
h) audit conclusions;
i) a statement on the degree to which the audit criteria have been fulfilled;
j) any unresolved diverging opinions between the audit team and the auditee;
k) audits by nature are a sampling exercise; as such there is a risk that the audit evidence examined is not representative.

The audit report can also include or refer to the following, as appropriate:

— the audit plan including time schedule;
— a summary of the audit process, including any obstacles encountered that may decrease the reliability of the audit conclusions;
— confirmation that the audit objectives have been achieved within the audit scope in accordance with the audit plan;
— any areas within the audit scope not covered

e)　監査活動を行った日時及び場所

f)　監査基準
g)　監査所見及び関連する証拠
h)　監査結論
i)　監査基準が満たされた程度に関する記述

j)　監査チームと被監査者との間で未解決の意見の相違
k)　監査とは，本質的にサンプリング作業であるということ．したがって，調査した監査証拠が代表的なものではないというリスクが存在する．

　監査報告書には，適宜，次の事項を含めるか，又はその事項の参照先を示すことができる．
— タイムスケジュールを含む監査計画
— 監査プロセスの要約．これには，監査結論の信頼性を低下させるかもしれない，監査中に遭遇した障害を含む．
— 監査計画に従って監査範囲内で監査目的を達成したことの確認

— 監査範囲内で監査しなかった領域．これには，

including any issues of availability of evidence, resources or confidentiality, with related justifications;
— a summary covering the audit conclusions and the main audit findings that support them;
— good practices identified;
— agreed action plan follow-up, if any;

— a statement of the confidential nature of the contents;
— any implications for the audit programme or subsequent audits.

6.5.2 Distributing audit report

The audit report should be issued within an agreed period of time. If it is delayed, the reasons should be communicated to the auditee and the individual(s) managing the audit programme.

The audit report should be dated, reviewed and accepted, as appropriate, in accordance with the audit programme.

The audit report should then be distributed to the

証拠の利用可能性，資源又は機密保持に関するあらゆる課題を，関係する根拠とともに含める．
— 監査結論及びそれを裏付ける主要な監査所見を含む概要
— 特定された優れた実践事例
— 存在する場合は，合意した処置の計画のフォローアップ
— 内容の機密性に関する記述

— 監査プログラム又はその後の監査に対する影響

6.5.2 監査報告書の配付

　監査報告書は，合意した期間内に発行することが望ましい．遅延する場合には，その理由を被監査者及び監査プログラムをマネジメントする人に連絡することが望ましい．

　監査報告書は，監査プログラムに従って，適切に，日付を付し，レビュー及び受諾することが望ましい．

　監査報告書は，次に，監査プログラム又は監査計

relevant interested parties defined in the audit programme or audit plan.

When distributing the audit report, appropriate measures to ensure confidentiality should be considered.

6.6 Completing audit

The audit is completed when all planned audit activities have been carried out, or as otherwise agreed with the audit client (e.g. there might be an unexpected situation that prevents the audit being completed according to the audit plan).

Documented information pertaining to the audit should be retained or disposed of by agreement between the participating parties and in accordance with audit programme and applicable requirements.

Unless required by law, the audit team and the individual(s) managing the audit programme should not disclose any information obtained during the audit, or the audit report, to any other par-

画で定めた関連する利害関係者へ配付することが望ましい．

監査報告書を配付する際は，機密保持を確実にするための適切な方策を考慮することが望ましい．

6.6 監査の完了

監査が完了するのは，全ての計画した監査活動を遂行したとき，又はそれ以外では監査依頼者と合意したときである（例えば，監査が，監査計画のとおりに完了することを妨げる予期しない事態があるであろう．）．

監査に関係する文書化した情報は，監査に参加した関係者間の合意によって，並びに監査プログラム及び適用される要求事項に従って，保持又は廃棄することが望ましい．

法律で要求されない限り，監査チーム及び監査プログラムをマネジメントする人は，監査依頼者の明確な承認なしに，及び被監査者の承認が必要な場合にそれなしに，監査中に入手したいかなる情報又は

ty without the explicit approval of the audit client and, where appropriate, the approval of the auditee. If disclosure of the contents of an audit document is required, the audit client and auditee should be informed as soon as possible.

Lessons learned from the audit can identify risks and opportunities for the audit programme and the auditee.

6.7 Conducting audit follow-up

The outcome of the audit can, depending on the audit objectives, indicate the need for corrections, or for corrective actions, or opportunities for improvement. Such actions are usually decided and undertaken by the auditee within an agreed timeframe. As appropriate, the auditee should keep the individual(s) managing the audit programme and/or the audit team informed of the status of these actions.

The completion and effectiveness of these actions should be verified. This verification may be part of a subsequent audit. Outcomes should be reported

監査報告書も,他の者に開示しないことが望ましい.監査文書の内容の開示を要求された場合は,できるだけ速やかに監査依頼者及び被監査者に知らせることが望ましい.

監査から得た知見から,監査プログラム及び被監査者に関するリスク及び機会を特定し得る.

6.7 監査のフォローアップの実施

監査の成果には,監査目的によって,修正若しくは是正処置の必要性,又は改善の機会を示すことができる.このような処置は,通常,合意した期間内に被監査者が決めて行う.適切な場合には,被監査者は,これらの処置の状況を,監査プログラムをマネジメントする人及び/又は監査チームに知らせておくことが望ましい.

これらの処置の完了及び有効性は,検証することが望ましい.この検証は,その後の監査の一部としてよい.成果は,監査プログラムをマネジメントす

to the individual managing the audit programme and reported to the audit client for management review.

7 Competence and evaluation of auditors
7.1 General

Confidence in the audit process and the ability to achieve its objectives depends on the competence of those individuals who are involved in performing audits, including auditors and audit team leaders. Competence should be evaluated regularly through a process that considers personal behaviour and the ability to apply the knowledge and skills gained through education, work experience, auditor training and audit experience. This process should take into consideration the needs of the audit programme and its objectives. Some of the knowledge and skills described in **7.2.3** are common to auditors of any management system discipline; others are specific to individual management system disciplines. It is not necessary for each auditor in the audit team to have the same competence. However, the overall competence of the audit team needs to be sufficient to achieve the audit objectives.

る人に報告し，マネジメントレビューのために監査依頼者へ報告することが望ましい．

7 監査員の力量及び評価
7.1 一般

　監査プロセス及びその目的を達成するための能力における信頼は，監査を行うことに関与する人々の力量に依存する．これらの人々には，監査員及び監査チームリーダーを含む．力量は，定期的に評価することが望ましい．この評価は，個人の行動，並びに教育，業務経験，監査員訓練及び監査経験によって身に付けた，知識及び技能を適用する能力を考慮するプロセスを通じて行う．このプロセスは，監査プログラム及びその目的のニーズを考慮に入れることが望ましい．**7.2.3** に示す知識及び技能には，あらゆるマネジメントシステム分野の監査員に共通のものもあれば，個々のマネジメントシステム分野の監査員に固有のものもある．監査チームにおける個々の監査員が同じ力量を備えている必要はない．ただし，監査チーム全体としての力量は，監査目的を達成するために十分である必要がある．

The evaluation of auditor competence should be planned, implemented and documented to provide an outcome that is objective, consistent, fair and reliable. The evaluation process should include four main steps, as follows:

a) determine the required competence to fulfil the needs of the audit programme;
b) establish the evaluation criteria;
c) select the appropriate evaluation method;
d) conduct the evaluation.

The outcome of the evaluation process should provide a basis for the following:

— selection of audit team members (as described in **5.5.4**);
— determining the need for improved competence (e.g. additional training);
— ongoing performance evaluation of auditors.

Auditors should develop, maintain and improve their competence through continual professional development and regular participation in audits (see **7.6**).

監査員の力量の評価は，計画し，実施し，文書化することが望ましい．これは，客観的で，一貫性をもち，公正で，かつ，信頼できる成果を提供するためである．この評価プロセスには，次の四つの主要なステップを含めることが望ましい．

a) 監査プログラムのニーズを満たすために必要な力量を決定する．
b) 評価基準を確立する．
c) 適切な評価方法を選択する．
d) 評価を行う．

評価プロセスの成果は，次の事項の基礎を提供することが望ましい．

— （**5.5.4** で示した）監査チームメンバーの選定

— 力量向上の必要性の決定（例えば，追加的な研修）
— 監査員の継続的なパフォーマンス評価

監査員は，専門能力の継続的開発及び監査への定期的な参加によって，自らの力量を開発し，維持し，向上することが望ましい（**7.6** 参照）．

A process for evaluating auditors and audit team leaders is described in **7.3**, **7.4** and **7.5**.

Auditors and audit team leaders should be evaluated against the criteria set out in **7.2.2** and **7.2.3** as well as the criteria established in **7.1**.

The competence required of the individual(s) managing the audit programme is described in **5.4.2**.

7.2 Determining auditor competence
7.2.1 General

In deciding the necessary competence for an audit, an auditor's knowledge and skills related to the following should be considered:

a) the size, nature, complexity, products, services and processes of auditees;

b) the methods for auditing;

c) the management system disciplines to be audited;

d) the complexity and processes of the management system to be audited;

e) the types and levels of risks and opportunities addressed by the management system;

監査員及び監査チームリーダーを評価するプロセスを **7.3**, **7.4** 及び **7.5** に示す.

監査員及び監査チームリーダーは，**7.1** で確立した基準だけでなく，**7.2.2** 及び **7.2.3** で設定した基準に対しても評価されることが望ましい.

監査プログラムをマネジメントする人に求められる力量を，**5.4.2** に示す.

7.2　監査員の力量の決定
7.2.1　一般
監査に求められる必要な力量を決めるときは，次の事項に関係する，監査員の知識及び技能を考慮することが望ましい.

a)　被監査者の規模，性質，複雑さ，製品，サービス及びプロセス
b)　監査の方法
c)　監査の対象となるマネジメントシステムの分野
d)　監査の対象となるマネジメントシステムの複雑さ及びプロセス
e)　マネジメントシステムで対処するリスク及び機会の，タイプ及びレベル

f) the objectives and extent of the audit programme;

g) the uncertainty in achieving audit objectives;

h) other requirements, such as those imposed by the audit client or other relevant interested parties, where appropriate.

This information should be matched against that listed in **7.2.3**.

7.2.2 Personal behaviour

Auditors should possess the necessary attributes to enable them to act in accordance with the principles of auditing as described in **Clause 4**. Auditors should exhibit professional behaviour during the performance of audit activities. Desired professional behaviours include being:

a) ethical, i.e. fair, truthful, sincere, honest and discreet;

b) open-minded, i.e. willing to consider alternative ideas or points of view;

c) diplomatic, i.e. tactful in dealing with individuals;

d) observant, i.e. actively observing physical sur-

f) 監査プログラムの目的及び監査プログラムの及ぶ領域
g) 監査目的の達成における不確かさ
h) 該当する場合,その他の要求事項.例えば,監査依頼者又はその他の関連する利害関係者によって課されるもの.

　この情報は,**7.2.3** に掲げる事項に対して合っていることが望ましい.

7.2.2　個人の行動

　監査員は,箇条 4 に示す監査の原則に従って活動するために必要な特質を備えていることが望ましい.監査員は,監査活動を実施している間,専門家としての行動を示すことが望ましい.望ましい専門家としての行動には,次の事項を含む.

a) 倫理的である.すなわち,公正である,信用できる,誠実である,正直である,そして分別がある.
b) 心が広い.すなわち,別の考え方又は視点を進んで考慮する.
c) 外交的である.すなわち,目的を達成するように人と上手に接する.
d) 観察力がある.すなわち,物理的な周囲の状況

roundings and activities;

e) perceptive, i.e. aware of and able to understand situations;

f) versatile, i.e. able to readily adapt to different situations;

g) tenacious, i.e. persistent and focused on achieving objectives;

h) decisive, i.e. able to reach timely conclusions based on logical reasoning and analysis;

i) self-reliant, i.e. able to act and function independently while interacting effectively with others;

j) able to act with fortitude, i.e. able to act responsibly and ethically, even though these actions may not always be popular and may sometimes result in disagreement or confrontation;

k) open to improvement, i.e. willing to learn from situations;

l) culturally sensitive, i.e. observant and respectful to the culture of the auditee;

m) collaborative, i.e. effectively interacting with others, including audit team members and the

7 監査員の力量及び評価

及び活動を積極的に観察する．

- e) 知覚が鋭い．すなわち，状況を認知し，理解できる．
- f) 適応性がある．すなわち，異なる状況に容易に合わせることができる．
- g) 粘り強い．すなわち，根気があり，目的の達成に集中する．
- h) 決断力がある．すなわち，論理的な理由付け及び分析に基づいて，時宜を得た結論に到達することができる．
- i) 自立的である．すなわち，他の人々と有効なやりとりをしながらも独立して活動し，役割を果たすことができる．
- j) 不屈の精神をもって活動できる．すなわち，その活動が，ときには受け入れられず，意見の相違又は対立をもたらすことがあっても，責任をもち，倫理的に活動することができる．

- k) 改善に対して前向きである．すなわち，進んで状況から学ぶ．
- l) 文化に対して敏感である．すなわち，被監査者の文化を観察し，尊重する．
- m) 協力的である．すなわち，監査チームメンバー及び被監査者の要員を含む他の人々とともに有

auditee's personnel.

7.2.3 Knowledge and skills
7.2.3.1 General
Auditors should possess:

a) the knowledge and skills necessary to achieve the intended results of the audits they are expected to perform;

b) generic competence and a level of discipline and sector-specific knowledge and skills.

Audit team leaders should have the additional knowledge and skills necessary to provide leadership to the audit team.

7.2.3.2 Generic knowledge and skills of management system auditors
Auditors should have knowledge and skills in the areas outlined below.

a) Audit principles, processes and methods: knowledge and skills in this area enable the auditor to ensure audits are performed in a consistent and systematic manner.

   An auditor should be able to:

7.2.3　知識及び技能
7.2.3.1　一般
監査員は,次の事項を備えていることが望ましい.

a)　実施が予定されている監査の，意図した結果を達成するのに必要な知識及び技能

b)　監査に共通に求められる力量，並びに分野及び業種に固有の知識及び技能のレベル

監査チームリーダーは，監査チームに対してリーダーシップを発揮するのに必要な付加的な知識及び技能を備えていることが望ましい.

7.2.3.2　マネジメントシステム監査員の共通的な知識及び技能
監査員は，次に概要を示す領域の知識及び技能を備えていることが望ましい.

a)　監査の原則，プロセス及び方法：この領域の知識及び技能によって，監査員は，一貫性のある体系的な監査の実施を確実にすることが可能となる.

　監査員は,次の事項ができることが望ましい.

- understand the types of risks and opportunities associated with auditing and the principles of the risk-based approach to auditing;
- plan and organize the work effectively;
- perform the audit within the agreed time schedule;
- prioritize and focus on matters of significance;
- communicate effectively, orally and in writing (either personally, or through the use of interpreters);
- collect information through effective interviewing, listening, observing and reviewing documented information, including records and data;
- understand the appropriateness and consequences of using sampling techniques for auditing;
- understand and consider technical experts' opinions;
- audit a process from start to finish, including the interrelations with other processes and different functions, where ap-

— 監査実施に付随するリスク及び機会のタイプ並びに監査実施へのリスクに基づくアプローチの原則を理解する．

— 有効に作業を計画し，必要な手配をする．
— 合意したタイムスケジュール内で監査を行う．
— 重要事項を優先し，重点的に取り組む．

— 口頭及び書面で有効にコミュニケーションを取る（自身で，又は通訳の利用を通じて）．

— 有効なインタビュー，聞き取り，観察，並びに記録及びデータを含む文書化した情報のレビューによって，情報を収集する．

— 監査のためにサンプリング技法を使用することの適切性及びそれによる結果を理解する．

— 技術専門家の意見を理解し，考慮する．

— 該当する場合，他のプロセス及び異なる機能との相互関係を含めて，プロセスを最初から最後まで監査する．

propriate;
- verify the relevance and accuracy of collected information;
- confirm the sufficiency and appropriateness of audit evidence to support audit findings and conclusions;
- assess those factors that may affect the reliability of the audit findings and conclusions;
- document audit activities and audit findings, and prepare reports;
- maintain the confidentiality and security of information.

b) Management system standards and other references: knowledge and skills in this area enable the auditor to understand the audit scope and apply audit criteria, and should cover the following:

- management system standards or other normative or guidance/supporting documents used to establish audit criteria or methods;
- the application of management system standards by the auditee and other organi-

- 収集した情報の関連性及び正確さを検証する．
- 監査所見及び監査結論の根拠とするために，監査証拠が十分かつ適切であることを確認する．
- 監査所見及び監査結論の信頼性に影響するかもしれない要因を評価する．
- 監査活動及び監査所見を文書化し，報告書を作成する．
- 情報の機密保持及びセキュリティを維持する．

b) マネジメントシステム規格及びその他の基準文書：この領域の知識及び技能によって，監査員は，監査範囲を理解し，監査基準を適用することが可能となる．この領域の知識及び技能には，次の事項を含めることが望ましい．
- 監査基準又は監査方法の確立に用いるマネジメントシステム規格又は他の規準文書若しくは手引・支援文書
- 被監査者及び他の組織によるマネジメントシステム規格の適用

zations;

- relationships and interactions between the management system(s) processes;
- understanding the importance and priority of multiple standards or references;
- application of standards or references to different audit situations.

c) The organization and its context: knowledge and skills in this area enable the auditor to understand the auditee's structure, purpose and management practices and should cover the following:

- needs and expectations of relevant interested parties that impact the management system;
- type of organization, governance, size, structure, functions and relationships;
- general business and management concepts, processes and related terminology, including planning, budgeting and management of individuals;
- cultural and social aspects of the auditee.

d) Applicable statutory and regulatory requirements and other requirements: knowledge and

- マネジメントシステムのプロセス間の関係及び相互作用
- 複数の規格又は基準文書の重要性及び優先順位の理解
- 様々な監査の位置づけへの規格又は基準文書の適用

c) **組織及び組織の状況**：この領域の知識及び技能によって，監査員は，被監査者の組織構造，目的及びそのマネジメントの実践を理解することが可能となる．この領域の知識及び技能には，次の事項を含めることが望ましい．

- マネジメントシステムに影響を及ぼす，関連する利害関係者のニーズ及び期待

- 組織のタイプ，統治，規模，構造，機能及び関係
- 全般的な事業及びそのマネジメントの概念，プロセス及び関係する用語．これには，計画，予算化及び人事管理を含む．

- 被監査者の文化的及び社会的側面

d) **適用される法令・規制要求事項及びその他の要求事項**：この領域の知識及び技能によって，監

skills in this area enable the auditor to be aware of, and work within, the organization's requirements. Knowledge and skills specific to the jurisdiction or to the auditee's activities, processes, products and services should cover the following:

— statutory and regulatory requirements and their governing agencies;
— basic legal terminology;
— contracting and liability.

> NOTE Awareness of statutory and regulatory requirements does not imply legal expertise and a management system audit should not be treated as a legal compliance audit.

7.2.3.3 Discipline and sector-specific competence of auditors

Audit teams should have the collective discipline and sector-specific competence appropriate for auditing the particular types of management systems and sectors.

査員は，組織の要求事項を認識すること，及びその枠内で監査業務を行うことが可能となる．法令，又は被監査者の活動，プロセス，製品，及びサービスに固有の知識及び技能には，次の事項を含めることが望ましい．

— 法令・規制要求事項及びその所管の行政機関

— 基本的な法的用語
— 契約及び法的責任

> 注記　法令・規制要求事項を認識しているということは，法律の専門家ということを意味しておらず，マネジメントシステム監査を法令順守の監査として扱うことは望ましくない．

7.2.3.3　分野及び業種に固有の監査員の力量

監査チームは，特定のタイプのマネジメントシステム及び業種を監査するのに適切な，その分野及び業種に固有の力量を<u>監査チーム</u>全体として備えていることが望ましい．

The discipline and sector-specific competence of auditors include the following:

a) management system requirements and principles, and their application;

b) fundamentals of the discipline(s) and sector(s) related to the management systems standards as applied by the auditee;

c) application of discipline and sector-specific methods, techniques, processes and practices to enable the audit team to assess conformity within the defined audit scope and generate appropriate audit findings and conclusions;

d) principles, methods and techniques relevant to the discipline and sector, such that the auditor can determine and evaluate the risks and opportunities associated with the audit objectives.

7.2.3.4 Generic competence of audit team leader

In order to facilitate the efficient and effective conduct of the audit an audit team leader should have the competence to:

a) plan the audit and assign audit tasks accord-

分野及び業種に固有の監査員の力量には，次の事項を含む．

a) マネジメントシステム要求事項及び原則，並びにそれらの適用

b) 被監査者が適用するマネジメントシステム規格に関係した，分野及び業種の基本

c) 分野及び業種に固有の方法，技法，プロセス，及び慣行の適用．これは，監査チームが定められた監査範囲内での適合性を評価し，適切な監査所見及び監査結論を導き出すことができるようにするためである．

d) 分野及び業種に関連した原則，方法及び技法．これは，監査員が監査目的に付随するリスク及び機会を決定及び評価できるようにする．

7.2.3.4　監査チームリーダーの共通的な力量

監査の効率的及び有効な実施を容易にするために，監査チームリーダーは，次の事項を行う力量を備えていることが望ましい．

a) 監査を計画し，個々の監査チームメンバーの固

ing to the specific competence of individual audit team members;

b) discuss strategic issues with top management of the auditee to determine whether they have considered these issues when evaluating their risks and opportunities;

c) develop and maintain a collaborative working relationship among the audit team members;

d) manage the audit process, including:

- making effective use of resources during the audit;
- managing the uncertainty of achieving audit objectives;
- protecting the health and safety of the audit team members during the audit, including ensuring compliance of the auditors with the relevant health and safety, and security arrangements;
- directing the audit team members;
- providing direction and guidance to auditors-in-training;
- preventing and resolving conflicts and

有の力量に応じて監査業務を割り当てる．

b) 被監査者のトップマネジメントと戦略的課題について意見交換する．これは，被監査者が組織として，そのリスク及び機会を評価する際にこれらの課題を考慮したかどうかを決定するためである．

c) 監査チームメンバー間に協力的な業務関係を構築し，維持する．

d) 次の事項を含む監査プロセスをマネジメントする．

— 監査中に資源を有効に利用する．

— 監査目的を達成することの不確かさをマネジメントする．

— 監査中の監査チームメンバーの安全衛生を保護する．これには，監査員が関連する安全衛生及びセキュリティに関する取決めの順守を確実にすることを含む．

— 監査チームメンバーを指揮する．

— 訓練中の監査員を指揮及び指導する．

— 必要な場合，監査チーム内のものを含めて，

problems that can occur during the audit, including those within the audit team, as necessary.

e) represent the audit team in communications with the individual(s) managing the audit programme, the audit client and the auditee;

f) lead the audit team to reach the audit conclusions;

g) prepare and complete the audit report.

7.2.3.5 Knowledge and skills for auditing multiple disciplines

When auditing multiple discipline management systems, the audit team member should have an understanding of the interactions and synergy between the different management systems.

Audit team leaders should understand the requirements of each of the management system standards being audited and recognize the limits of their competence in each of the disciplines.

NOTE Audits of multiple disciplines done simultaneously can be done as a combined audit or as an

監査中に発生し得る利害抵触及び問題を防ぎ,解決する.

e) 監査プログラムをマネジメントする人,監査依頼者及び被監査者とのコミュニケーションでは監査チームを代表する.
f) 監査チームを導いて,監査結論に達する.

g) 監査報告書を作成し,完成する.

7.2.3.5 複数分野を監査するための知識及び技能

複数分野のマネジメントシステムを監査する際は,監査チームメンバーは異なるマネジメントシステム間の相互作用及び相乗効果を理解していることが望ましい.

監査チームリーダーは,監査対象となっている各マネジメントシステム規格の要求事項を理解し,それぞれの分野における自身の力量の限界を認識することが望ましい.

> 注記　複数分野を同時に監査することは,複合監査として又は複数分野を含む統合マネ

audit of an integrated management system that covers multiple disciplines.

7.2.4 Achieving auditor competence

Auditor competence can be acquired using a combination of the following:

a) successfully completing training programmes that cover generic auditor knowledge and skills;

b) experience in a relevant technical, managerial or professional position involving the exercise of judgement, decision making, problem solving and communication with managers, professionals, peers, customers and other relevant interested parties;

c) education/training and experience in a specific management system discipline and sector that contribute to the development of overall competence;

d) audit experience acquired under the supervision of an auditor competent in the same discipline.

NOTE   Successful completion of a training course

ジメントシステムの監査として行うことが可能である．

7.2.4 監査員の力量の獲得

監査員の力量は，次の組合せによって獲得し得る．

a) 共通的な監査員の知識及び技能を対象とする訓練プログラムの成功裏の完了

b) 関連する技術的，管理的又は専門的職位での経験．これは，判断の行使，意思決定，問題解決，並びに管理者，専門家，同僚，顧客及びその他の関連する利害関係者とのコミュニケーションに関与するものである．

c) 全体としての力量の開発に寄与する，特定のマネジメントシステムの分野及び業種についての教育・訓練及び経験

d) 同じ分野で力量のある監査員の監督下で獲得する監査経験

   注記 訓練コースの成功裏の完了かどうかは，

will depend on the type of course. For courses with an examination component it can mean successfully passing the examination. For other courses, it can mean participating in and completing the course.

7.2.5 Achieving audit team leader competence

An audit team leader should have acquired additional audit experience to develop the competence described in **7.2.3.4**. This additional experience should have been gained by working under the direction and guidance of a different audit team leader.

7.3 Establishing auditor evaluation criteria

The criteria should be qualitative (such as having demonstrated desired behaviour, knowledge or the performance of the skills, in training or in the workplace) and quantitative (such as the years of work experience and education, number of audits conducted, hours of audit training).

7.4 Selecting appropriate auditor evalua-

そのコースのタイプに依存するであろう．試験を含むコースでは試験に合格することを意味し得る．他のコースではコースに参加し，完了することを意味し得る．

7.2.5 監査チームリーダーの力量の獲得

監査チームリーダーは，7.2.3.4 に示す力量を開発するための追加の監査経験を獲得していることが望ましい．この追加の経験は，他の監査チームリーダーの指揮及び指導の下での監査業務によって得られたものであることが望ましい．

7.3 監査員の評価基準の確立

この基準には，定性的（例えば，訓練又は職場で示された，望ましい行動，知識又は技能のパフォーマンス）及び定量的（例えば，業務経験及び教育の年数，監査を行った回数，監査員研修の時間）なものがあることが望ましい．

7.4 監査員の適切な評価方法の選択

tion method

The evaluation should be conducted using two or more of the methods given in **Table 2**. In using **Table 2**, the following should be noted:

a) the methods outlined represent a range of options and may not apply in all situations;

b) the various methods outlined may differ in their reliability;

c) a combination of methods should be used to ensure an outcome that is objective, consistent, fair and reliable.

Table 2 — Auditor evaluation methods

Evaluation method	Objectives	Examples
Review of records	To verify the background of the auditor	Analysis of records of education, training, employment, professional credentials and auditing experience
Feedback	To provide information about how the performance of the auditor is perceived	Surveys, questionnaires, personal references, testimonials, complaints, performance evaluation, peer review

7 監査員の力量及び評価

評価は,表2に示す方法の二つ以上を利用して行うことが望ましい.表2を利用するときは,次の事項に注意することが望ましい.

a) 表2に概要を示した方法は,様々な選択肢の中の代表的なものであり,全ての状況に適用してよいとは限らない.

b) 表2に概要を示した様々な方法の信頼性は,それぞれ異なってよい.

c) 評価結果が客観的で,一貫性をもち,公正で,かつ,信頼できることを確実にするために,複数の評価方法を組み合わせて用いることが望ましい.

表2 −監査員の評価方法

評価方法	目的	例
記録のレビュー	監査員の経歴を検証する.	教育,訓練,雇用,専門家としての資格及び監査経験の記録の解析
フィードバック	監査員のパフォーマンスがどのように受け止められているかに関する情報を与える.	調査,質問票,推薦状,お礼状,苦情,パフォーマンス評価,相互評価

Table 2 *(continued)*

Evaluation method	Objectives	Examples
Interview	To evaluate desired professional behaviour and communication skills, to verify information and test knowledge and to acquire additional information	Personal interviews
Observation	To evaluate desired professional behaviour and the ability to apply knowledge and skills	Role playing, witnessed audits, on-the-job performance
Testing	To evaluate desired behaviour and knowledge and skills and their application	Oral and written exams, psychometric testing
Post-audit review	To provide information on the auditor performance during the audit activities, identify strengths and opportunities for improvement	Review of the audit report, interviews with the audit team leader, the audit team and, if appropriate, feedback from the auditee

7.5 Conducting auditor evaluation

The information collected about the auditor under evaluation should be compared against the criteria set in **7.2.3**. When an auditor under evaluation who is expected to participate in the audit programme does not fulfil the criteria, then additional training, work or audit experience should be undertaken and a subsequent re-evaluation should be performed.

7 監査員の力量及び評価

表 2（続き）

評価方法	目的	例
インタビュー	望ましい専門家としての行動及びコミュニケーションの技能を評価し，情報を検証し，知識を試験し，並びに追加情報を獲得する．	個人面談
観察	望ましい専門家としての行動，並びに知識及び技能を適用する能力を評価する．	ロールプレイ，立会い監査，監査業務中のパフォーマンス
試験	望ましい行動，並びに知識，技能及びそれらの適用を評価する．	口頭及び筆記試験，心理試験
監査後のレビュー	監査活動中の監査員のパフォーマンスに関する情報を与え，強み及び改善の機会を特定する．	監査報告書のレビュー，監査チームリーダー，監査チームへのインタビュー，適切な場合は被監査者からのフィードバック

7.5 監査員の評価の実施

評価対象の監査員について収集した情報を 7.2.3 で設定した基準と比較することが望ましい．評価対象の監査員が，監査プログラムに参加が見込まれていて，評価基準を満たさないときには，追加の訓練，業務経験又は監査経験を積ませ，それに続く再評価を行うことが望ましい．

7.6 Maintaining and improving auditor competence

Auditors and audit team leaders should continually improve their competence. Auditors should maintain their auditing competence through regular participation in management system audits and continual professional development. This may be achieved through means such as additional work experience, training, private study, coaching, attendance at meetings, seminars and conferences or other relevant activities.

The individual(s) managing the audit programme should establish suitable mechanisms for the continual evaluation of the performance of the auditors and audit team leaders.

The continual professional development activities should take into account the following:

a) changes in the needs of the individual and the organization responsible for the conduct of the audit;

b) developments in the practice of auditing including the use of technology;

7.6 監査員の力量の維持及び向上

　監査員及び監査チームリーダーは，継続的にその力量を向上することが望ましい．監査員は，マネジメントシステムの監査への定期的な参加及び専門能力の継続的開発によって，監査の力量を維持することが望ましい．これは，次のような手段で達成してよい．例えば，追加の業務経験，訓練，個人学習，業務指導並びに会合，セミナー及び会議への参加，又はその他関連する諸活動がある．

　監査プログラムをマネジメントする人は，監査員及び監査チームリーダーのパフォーマンスの継続的評価のための適切な仕組みを確立することが望ましい．

　専門能力の継続的開発活動では，次の事項を考慮に入れることが望ましい．
a) 　監査の実施に責任をもつ個人及び組織の，ニーズの変化

b) 　技術の利用を含む，監査の実践における開発

c) relevant standards including guidance/supporting documents and other requirements;
d) changes in sector or disciplines.

c) 手引・支援文書を含む関連する規格,及びその他の要求事項
d) 業種又は分野における変化

Annex A

(informative)

Additional guidance for auditors planning and conducting audits

A.1 Applying audit methods

An audit can be performed using a range of audit methods. An explanation of commonly used audit methods can be found in this annex. The audit methods chosen for an audit depend on the defined audit objectives, scope and criteria, as well as duration and location. Available auditor competence and any uncertainty arising from the application of audit methods should also be considered. Applying a variety and combination of different audit methods can optimize the efficiency and effectiveness of the audit process and its outcome.

Performance of an audit involves an interaction among individuals within the management system being audited and the technology used to conduct the audit. **Table A.1** provides examples of audit methods that can be used, singly or in combination, in order to achieve the audit objectives. If an

附属書 A

(参考)

監査を計画及び実施する監査員に対する追加の手引

A.1 監査方法の適用

監査は多様な監査方法を利用して実行し得る．一般的に利用される監査方法の説明が，この附属書に見出し得る．監査に対して選ぶ監査方法は，定められた監査の目的，範囲及び基準，並びに期間及び場所による．利用可能な監査員の力量，及び監査方法の適用に起因する不確かさも考慮することが望ましい．異なった監査方法を様々に組み合わせて適用することによって，監査プロセス及びその成果の効率及び有効性を最適化することができる．

監査のパフォーマンスは，監査の対象となっているマネジメントシステムにおける個々人の相互作用及び監査の実施で利用する技術に関与するものである．**表 A.1** は，監査方法の例を提示し，これは，監査目的を達成するために，単独に，又は組み合わせて利用し得る．もし監査で複数メンバーの監査チ

audit involves the use of an audit team with multiple members, both on-site and remote methods may be used simultaneously.

NOTE   Additional information on visiting physical locations is given in **A.15**.

Table A.1 — **Audit methods**

Extent of involvement between the auditor and the auditee	Location of the auditor	
	On-site	Remote
Human interaction	Conducting interviews Completing checklists and questionnaires with auditee participation Conducting document review with auditee participation Sampling	Via interactive communication means: — conducting interviews; — observing work performed with remote guide; — completing checklists and questionnaires; — conducting document review with auditee participation.
No human interaction	Conducting document review (e.g. records, data analysis) Observing work performed Conducting on-site visit Completing checklists Sampling (e.g. products)	Conducting document review (e.g. records, data analysis) Observing work performed via surveillance means, considering social and statutory and regulatory requirements Analysing data

ームを使うのであれば,現地監査及び遠隔監査の両方の方法を同時に利用してもよい.

注記 物理的場所の訪問についての追加的な情報を **A.15** に示す.

表 A.1 − 監査方法

監査員と被監査者との間の関わりの程度	監査員の活動場所	
	現地	遠隔
人的交流あり	インタビューを行う. 被監査者の参加の下にチェックリスト及び質問事項を漏れなく確認する. 被監査者の参加の下に文書レビューを行う. サンプリングする.	双方向のコミュニケーション手段を通じて,次の事項を実施する. — インタビューを行う. — 遠隔監査の案内役とともに,実行中の作業を観察する. — チェックリスト及び質問事項を漏れなく確認する. — 被監査者の参加の下に文書レビューを行う.
人的交流なし	文書レビューを行う(例えば,記録,データ分析). 実行中の作業を観察する. 現地訪問を行う. チェックリストを漏れなく確認する. サンプリング(例えば,製品)する.	文書レビューを行う(例えば,記録,データ分析). 社会的要求事項及び法令・規制要求事項を考慮しながら,サーベイランス手段を通じて,実行中の作業を観察する. データを分析する.

Table A.1 *(continued)*

> On-site audit activities are performed at the location of the auditee. Remote audit activities are performed at any place other than the location of the auditee, regardless of the distance.
>
> Interactive audit activities involve interaction between the auditee's personnel and the audit team. Non-interactive audit activities involve no human interaction with individuals representing the auditee but do involve interaction with equipment, facilities and documentation.

The responsibility of the effective application of audit methods for any given audit in the planning stage remains with either the individual(s) managing the audit programme or the audit team leader. The audit team leader has this responsibility for conducting the audit activities.

The feasibility of remote audit activities can depend on several factors (e.g. the level of risk to achieving the audit objectives, the level of confidence between auditor and auditee's personnel and regulatory requirements).

At the level of the audit programme, it should be ensured that the use of remote and on-site application of audit methods is suitable and balanced, in order to ensure satisfactory achievement of audit

表 A.1（続き）

> 現地監査活動は，被監査者の場所で行われる．遠隔監査活動は，距離とは無関係に，被監査者の場所以外のあらゆる場所で行われる．
>
> 対話的な監査活動は，被監査者の要員と監査チームとの間の人的交流を含む．対話的でない監査活動は，被監査者を代表する人々との人的交流を含まないが，機器，施設及び文書類との関わりを含む．

　計画策定段階で，所与の監査に監査方法を有効に適用することの責任は，監査プログラムをマネジメントする人又は監査チームリーダーにある．監査チームリーダーは，この，監査活動の実施に対する責任をもつ．

　遠隔監査活動の実施可能性は，幾つかの要因（例えば，監査目的の達成に対するリスクのレベル，監査員と被監査者の要員との間の信頼のレベル，規制要求事項）に基づき得る．

　監査プログラムのレベルで，監査方法に関して遠隔監査及び現地監査を，適切に，かつ，バランスよく適用して使うことを確実にすることが望ましい．これは，監査プログラムの目的の達成を満たすこと

programme objectives.

A.2 Process approach to auditing

The use of a "process approach" is a requirement for all ISO management system standards in accordance with ISO/IEC Directives, Part 1, Annex SL. Auditors should understand that auditing a management system is auditing an organization's processes and their interactions in relation to one or more management system standard(s). Consistent and predictable results are achieved more effectively and efficiently when activities are understood and managed as interrelated processes that function as a coherent system.

A.3 Professional judgement

Auditors should apply professional judgement during the audit process and avoid concentrating on the specific requirements of each clause of the standard at the expense of achieving the intended outcome of the management system. Some ISO management system standard clauses do not readily lend themselves to audit in terms of comparison between a set of criteria and the content of a proce-

を確実にするためである．

A.2　監査に対するプロセスアプローチ

"プロセスアプローチ"の利用は，**ISO/IEC** 専門業務用指針第 1 部の**附属書 SL** に従う，全ての **ISO** マネジメントシステム規格における要求事項である．マネジメントシステム監査とは，一つ又は複数のマネジメントシステム規格に関係する組織のプロセス及びそれらの相互作用を監査することであることを，監査員は理解することが望ましい．一貫して予測可能な結果を，より有効にかつ効率的に達成するのは，マネジメントシステムの活動を，首尾一貫したシステムとして機能する相互に関連したプロセスとして理解し，マネジメントするときである．

A.3　専門的な判断

監査員は，監査プロセスにおいて専門的な判断を適用することが望ましい．その際，マネジメントシステムの意図した成果の達成を見る観点を犠牲の上で，規格の各箇条の個別の要求事項への集中をするような監査は避けることが望ましい．ISO マネジメントシステム規格の箇条の中には，一連の基準として，手順又は作業指示の内容との比較の観点での監査には，その箇条そのものだけではなじまないも

dure or work instruction. In these situations, auditors should use their professional judgement to determine whether the intent of the clause has been met.

A.4 Performance results

Auditors should be focused on the intended result of the management system throughout the audit process. While processes and what they achieve are important, the result of the management system and its performance are what counts. It is also important to consider the level of the integration of different management systems and their intended results.

The absence of a process or documentation can be important in a high risk or complex organization but not so significant in other organizations.

A.5 Verifying information

Insofar as practicable, the auditors should consider whether the information provides sufficient objective evidence to demonstrate that requirements are being met, such as being:

のもある．このような状況では，監査員は，その箇条の意図が満たされているかどうかを決定するために，専門的な判断を用いることが望ましい．

A.4 パフォーマンスの結果

監査員は，監査プロセス全体を通して，そのマネジメントシステムの意図した結果に焦点を当てることが望ましい．プロセス及びそれらが達成するものは重要ではあるが，肝心なことはマネジメントシステムの結果及びマネジメントシステムのパフォーマンスである．また，異なるマネジメントシステムの統合のレベル及びそれらの意図した結果を考慮することも重要である．

プロセス又は文書類の欠如は，リスクの高い又は複雑な組織においては重要であり得るが，他の組織においてはそれほど重要ではないこともあり得る．

A.5 情報の検証

実行可能である限り，監査員は，情報が，要求事項を満たしていることを実証するのに十分な客観的証拠を提供するものであるかどうか，例えば，次の事項を考慮することが望ましい．

a) complete (all expected content is contained in the documented information);
b) correct (the content conforms to other reliable sources such as standards and regulations);
c) consistent (the documented information is consistent in itself and with related documents);
d) current (the content is up to date).

It should also be considered whether the information being verified provides sufficient objective evidence to demonstrate that requirements are being met.

If information is provided in a manner other than expected (e.g. by different individuals, alternate media), the integrity of the evidence should be assessed.

Specific care is needed for information security due to applicable regulations on protection of data (in particular for information which lies outside the audit scope, but which is also contained in the document).

a) 完全である（全ての期待される内容が文書化した情報に含まれている．）．
b) 適正である（内容が規格及び規制のような他の信頼できる情報源に適合している．）．
c) 一貫している（文書化した情報が，それ自体で一貫している，及び関係する文書との一貫性がある．）．
d) 現行のものである（内容が更新されている）．

　検証中の情報が，要求事項を満たしていることを実証するのに十分な客観的証拠を提供するものであるかどうかも考慮することが望ましい．

　情報が，予期したものと異なる方法で（例えば，異なる人々，代わりの媒体によって）提供されるならば，その証拠についての完全性を評価することが望ましい．

　特別な注意が情報セキュリティに対して必要となるのは，データ（特に情報として，監査範囲外におかれているが，文書に含まれもしている）の保護に関する適用可能な規制によるものである．

A.6 Sampling

A.6.1 General

Audit sampling takes place when it is not practical or cost effective to examine all available information during an audit, e.g. records are too numerous or too dispersed geographically to justify the examination of every item in the population. Audit sampling of a large population is the process of selecting less than 100 % of the items within the total available data set (population) to obtain and evaluate evidence about some characteristic of that population, in order to form a conclusion concerning the population.

The objective of audit sampling is to provide information for the auditor to have confidence that the audit objectives can or will be achieved.

The risk associated with sampling is that the samples may not be representative of the population from which they are selected. Thus, the auditor's conclusion may be biased and be different from that which would be reached if the whole population was examined. There may be other risks de-

A.6 サンプリング
A.6.1 一般

監査サンプリングは,監査において全ての利用可能な情報を調査するのが現実的ではない場合,又は費用対効果が高くない場合,例えば,記録が,母集団内の全ての対象項目の調査を正当化するにはあまりに膨大であったり,地理的にあまりに分散していたりする場合に行う.大きな母集団からの監査サンプリングは,母集団のある種の特性について証拠を得て評価するために,利用可能なデータセット全体(母集団)の中から,対象項目の 100% 未満の項目を選択するプロセスである.これは,その母集団に関する結論を形成するためである.

監査サンプリングの目的は,監査員に,監査目的を達成できる又は達成するであろうという確信をもてる情報を提供することである.

サンプリングに付随するリスクは,サンプルがその選択元である母集団を代表していない場合があることである.その結果,監査員の結論が偏り,母集団の全てを調査した際に達するであろう結論とは異なる場合がある.他のリスクがある場合としては,サンプルする母集団におけるばらつき及び選択した

pending on the variability within the population to be sampled and the method chosen.

Audit sampling typically involves the following steps:
a) establishing the objectives of sampling;
b) selecting the extent and composition of the population to be sampled;
c) selecting a sampling method;
d) determining the sample size to be taken;
e) conducting the sampling activity;
f) compiling, evaluating, reporting and documenting results.

When sampling, consideration should be given to the quality of the available data, as sampling insufficient and inaccurate data will not provide a useful result. The selection of an appropriate sample should be based on both the sampling method and the type of data required, e.g. to infer a particular behaviour pattern or draw inferences across a population.

Reporting on the sample selected could take into

サンプリング方法によるものがある．

監査サンプリングは一般的に次のステップを含む．
a) サンプリングの目的を確立する．
b) サンプルする母集団の範囲及び構成を選択する．
c) サンプリング方法を選択する．
d) 採取するサンプルサイズを決定する．
e) サンプリング活動を行う．
f) 結果をまとめ，評価，報告及び文書化する．

サンプリングするとき，利用可能なデータの品質を考慮することが望ましい．これは，サンプリングのデータが不十分で不正確だと，有用な結果とならないからである．適切なサンプルの選択は，サンプリング方法及び求められるデータのタイプの両方に基づくことが望ましい．例えば，母集団全体にわたって，特定の振舞いのパターンを推測するため，又は全体にわたる推測を引き出すためである．

選択したサンプルについての報告は，そのサンプ

account the sample size, selection method and estimates made based on the sample and the confidence level.

Audits can use either judgement-based sampling (see **A.6.2**) or statistical sampling (see **A.6.3**).

A.6.2 Judgement-based sampling

Judgement-based sampling relies on the competence and experience of the audit team (see **Clause 7**).

For judgement-based sampling, the following can be considered:

a) previous audit experience within the audit scope;
b) complexity of requirements (including statutory and regulatory requirements) to achieve the audit objectives;
c) complexity and interaction of the organization's processes and management system elements;
d) degree of change in technology, human factor

ルサイズ,選択の方法,並びにサンプルに基づいた推定及びその信頼水準を考慮に入れ得るであろう.

監査は,判断に基づくサンプリング(**A.6.2** 参照)又は統計的サンプリング(**A.6.3** 参照)のいずれかを使うことができる.

A.6.2 判断に基づくサンプリング

判断に基づくサンプリングは,監査チームの力量及び経験に依存する(箇条 **7** 参照).

判断に基づくサンプリングに対して,次の事項を考慮し得る.

a) 監査範囲内の前回までの監査経験

b) 監査目的を達成するための要求事項(法令・規制要求事項を含む.)の複雑さ

c) 組織のプロセス及びマネジメントシステム要素の複雑さ及び相互作用

d) 技術,人的要因又はマネジメントシステムの変

or management system;

e) previously identified significant risks and opportunities for improvement;

f) output from monitoring of management systems.

A drawback to judgement-based sampling is that there can be no statistical estimate of the effect of uncertainty in the findings of the audit and the conclusions reached.

A.6.3 Statistical sampling

If the decision is made to use statistical sampling, the sampling plan should be based on the audit objectives and what is known about the characteristics of overall population from which the samples are to be taken.

Statistical sampling design uses a sample selection process based on probability theory. Attribute-based sampling is used when there are only two possible sample outcomes for each sample (e.g. correct/incorrect or pass/fail). Variable-based sampling is used when the sample outcomes occur in a

化の度合い

e) 以前に特定された重要なリスク及び改善の機会

f) マネジメントシステムの監視からのアウトプット

　判断に基づくサンプリングの欠点は，監査の所見及びその達した結論における，不確かさの影響に関する統計的な推定が存在し得ないことである．

A.6.3 統計的サンプリング

　統計的サンプリングを使うことに決めるならば，サンプリング計画は，監査目的，及びサンプル採取対象の母集団の全体としての特性に関して知られていることに基づくことが望ましい．

　統計的サンプリングの設計には，確率論に基づいたサンプル選択プロセスを使う．属性に基づくサンプリングを使うのは，各サンプルに対するサンプル結果に二つの可能性（例えば，正か誤，合か否）しかない場合である．変数に基づくサンプリングを使うのは，サンプル結果がある連続した範囲において

continuous range.

The sampling plan should take into account whether the outcomes being examined are likely to be attribute-based or variable-based. For example, when evaluating conformity of completed forms to the requirements set out in a procedure, an attribute-based approach could be used. When examining the occurrence of food safety incidents or the number of security breaches, a variable-based approach would likely be more appropriate.

Elements that can affect the audit sampling plan are:
a) the context, size, nature and complexity of the organization;
b) the number of competent auditors;
c) the frequency of audits;
d) the time of individual audit;
e) any externally required confidence level;
f) the occurrence of undesirable and/or unexpected events.

When a statistical sampling plan is developed, the

発生する場合である．

　サンプリング計画は，調査される結果が属性に基づくものになりやすいか，変数に基づくものになりやすいかを考慮に入れることが望ましい．例えば，手順の中で設定された要求事項に対して，記入済みの書式の適合性を評価する場合は，属性に基づくアプローチを使い得るであろう．食品安全に関するインシデントの発生又はセキュリティの漏えい（洩）の数を調査する場合は，変数に基づくアプローチがより適切となりやすいだろう．

　監査サンプリング計画に影響を及ぼし得る要素は，次の事項である．
a)　組織の状況，規模，性質及び複雑さ

b)　力量のある監査員の数
c)　監査の頻度
d)　個々の監査の工数
e)　外部から求められる信頼水準
f)　望ましくない及び／又は予期しない事象の発生

　統計的サンプリング計画を策定する場合，どのレ

level of sampling risk that the auditor is willing to accept is an important consideration. This is often referred to as the acceptable confidence level. For example, a sampling risk of 5 % corresponds to an acceptable confidence level of 95 %. A sampling risk of 5 % means the auditor is willing to accept the risk that 5 out of 100 (or 1 in 20) of the samples examined will not reflect the actual values that would be seen if the entire population was examined.

When statistical sampling is used, auditors should appropriately document the work performed. This should include a description of the population that was intended to be sampled, the sampling criteria used for the evaluation (e.g. what is an acceptable sample), the statistical parameters and methods that were utilized, the number of samples evaluated and the results obtained.

A.7 Auditing compliance within a management system

The audit team should consider if the auditee has effective processes for:

ベルのサンプリングリスクならば監査員が受容しようとするかは，重要な考慮事項である．これは，受容可能な信頼水準と呼ばれることが多い．例えば，5%のサンプリングリスクは95%の受容可能な信頼水準に相当する．5%のサンプリングリスクは，次のようなリスクならば監査員が受容することを意味する．"調査したサンプルの100のうち5（又は20のうち1）は，全体の母集団を調査すれば見られる実際の値を反映していない"というリスク．

統計的サンプリングを使う場合，監査員は実施した作業を適切に文書化することが望ましい．これには，サンプルすることを意図した母集団に関する記述，評価に使用したサンプリング基準（例えば，受容可能なサンプルとは何か），利用した統計的指標及び方法，評価したサンプルの数並びに取得した結果を含むことが望ましい．

A.7 マネジメントシステムにおける順守の監査

監査チームは，被監査者が，次の事項に対して有効なプロセスをもっているかどうかについて考慮す

a) identifying its statutory and regulatory requirements and other requirements it is committed to;
b) managing its activities, products and services to achieve compliance with these requirements;
c) evaluating its compliance status.

In addition to the generic guidance given in this document, when assessing the processes that the auditee has implemented to ensure compliance with relevant requirements, the audit team should consider if the auditee:

1) has an effective process for identifying changes in compliance requirements and for considering them as part of the management of change;
2) has competent individuals to manage its compliance processes;
3) maintains and provides appropriate documented information on its compliance status as required by regulators or other interested parties;

ることが望ましい.
a) 被監査者に適用される法令・規制要求事項,及び被監査者がコミットメントするその他の要求事項を特定する.
b) これらの要求事項への順守を達成するために,被監査者の活動,製品及びサービスをマネジメントする.
c) 被監査者の順守の状態を評価する.

この規格に示す共通的な手引に加えて,関連する要求事項への順守を確実にするために被監査者が実施したプロセスを監査で評価する場合,監査チームは,被監査者が,次の事項を行っているかを考慮することが望ましい.
1) 順守の要求事項における変更を特定し,これを変更のマネジメントの一環として考慮するための有効なプロセスをもつ.

2) 被監査者の順守のプロセスをマネジメントする力量を備えた人々をもつ.
3) 規制当局又は他の利害関係者から要求されているように,被監査者の順守の状態に関する,適切な文書化した情報を維持し,提供する.

4) includes compliance requirements in its internal audit programme;
5) addresses any instances of non-compliance;
6) considers compliance performance in its management reviews.

A.8 Auditing context

Many management systems standards require an organization to determine its context, including the needs and expectations of relevant interested parties and external and internal issues. To do this, an organization can use various techniques for strategic analysis and planning.

Auditors should confirm that suitable processes have been developed for this and are used effectively, so that their results provide a reliable basis for determining the scope and the development of the management system. To do this, auditors should consider objective evidence related to the following:

a) the process(es) or method(s) used;
b) the suitability and competence of the individuals contributing to the process(es);

4) 順守の要求事項を,被監査者の内部監査プログラムに含める.
5) 不順守のあらゆる事例に対処する.
6) 被監査者のマネジメントレビューにおいて,順守のパフォーマンスを考慮する.

A.8 組織の状況の監査

多くのマネジメントシステム規格は,組織が,その組織の状況を決定することを要求している.組織の状況には,関連する利害関係者のニーズ及び期待並びに外部及び内部の課題を含める.組織の状況を決定するために,組織は,戦略的な分析及び計画策定について様々な技法を使用し得る.

監査員は,このために適切なプロセス群が策定されており,有効に使用されていることを確認することが望ましい.それは,それらのプロセスの諸結果が,マネジメントシステムの適用範囲及び策定を決定するための信頼できる基礎を提供するからである.これを確認するために,監査員は,次の事項に関係する客観的証拠を考慮することが望ましい.

a) 使用したプロセス(群)又は方法(類)
b) 使用したプロセス(群)に寄与している個々人の適切性及び力量

c) the results of the process(es);
d) the application of the results to determine management system scope and development;

e) periodic reviews of context, as appropriate.

Auditors should have relevant sector-specific knowledge and understanding of the management tools that organizations can use in order to make a judgement regarding the effectiveness of the processes used to determine context.

A.9 Auditing leadership and commitment

Many management systems standards have increased requirements for top management.

These requirements include demonstrating commitment and leadership by taking accountability for the effectiveness of the management system and fulfilling a number of responsibilities. These include tasks that top management should undertake itself and others that can be delegated.

c)   使用したプロセス(群)の諸結果
d)   マネジメントシステムの適用範囲及び策定を決定するための,使用したプロセス(群)の諸結果の適用
e)   必要な場合,組織の状況の定期的レビュー

　監査員は,関連する業種に固有の知識,及び組織が使用し得るマネジメントツール類に関する理解をもつことが望ましい.これは,監査員が,組織の状況を決定するのに使用されたプロセス群の有効性について判断するためである.

A.9　リーダーシップ及びコミットメントの監査

　多くのマネジメントシステム規格は,トップマネジメントに対する要求事項を増やしてきた.

　これらの要求事項には,トップマネジメントがマネジメントシステムの有効性に対して説明責任(accountability)を負い,幾つかの責任を果たすことによって,コミットメント及びリーダーシップを実証することを含む.これらには,トップマネジメントが自身で実施することが望ましい業務及び他の者に委任し得る業務を含む.

Auditors should obtain objective evidence of the degree to which top management is involved in decision-making related to the management system and how it demonstrates commitment to ensuring its effectiveness. This can be achieved by reviewing the results from relevant processes (for example policies, objectives, available resources, communications from top management) and by interviewing staff to determine the degree of top management engagement.

Auditors should also aim to interview top management to confirm that they have an adequate understanding of the discipline-specific issues relevant to their management system, together with the context their organization operates within, so that they can ensure that the management system achieves its intended results.

Auditors should not only focus on leadership at the top management level but should also audit leadership and commitment at other levels of manage-

監査員は，トップマネジメントが，マネジメントシステムに関係する意思決定に参画している程度，及びマネジメントシステムの有効性を確実にすることへのコミットメントをどのようにして実証しているかについて，客観的証拠を得ることが望ましい．これは，関連プロセスからの結果［例えば，方針，目的（又は目標），利用可能な資源，トップマネジメントからのコミュニケーション］をレビューし，トップマネジメントの積極的参加の程度を決定するためにスタッフにインタビューすることによって達成できる．

監査員はまた，次の事項を確認することをトップマネジメントへのインタビューの目指すこととするのが望ましい．すなわち，トップマネジメントが，マネジメントシステムが意図した結果を達成することを確実にできるようにするために，マネジメントシステムに関連する分野固有の課題を，組織が運用する状況とともに十分に理解しているかということである．

監査員は，トップマネジメントレベルにおけるリーダーシップに焦点を当てるだけではなく，必要に応じて，マネジメントのその他のレベルにおけるリ

ment, as appropriate.

A.10 Auditing risks and opportunities

As part of the assignment of an individual audit the determination and management of the organization's risk and opportunities can be included. The core objectives for such an audit assignment are to:

— give assurance on the credibility of the risk and opportunity identification process(es);
— give assurance that risks and opportunities are correctly determined and managed;
— review how the organization addresses its determined risks and opportunities.

An audit of an organization's approach to the determination of risks and opportunities should not be performed as a stand-alone activity. It should be implicit during the entire audit of a management system, including when interviewing top management. An auditor should act in accordance with the following steps and collect objective evidence as follows:

ーダーシップ及びコミットメントを監査することもまた望ましい.

A.10　リスク及び機会の監査

　個々の監査の割当ての一部として,被監査者の組織のリスク及び機会の決定及びマネジメントを含み得る.このような監査の割当ての主たる目的は,次の事項のとおりである.

— リスク及び機会を特定するプロセス(群)の信頼性について保証を与える.
— リスク及び機会を適正に決定してマネジメントすることに保証を与える.
— 組織が,その決定したリスク及び機会にどのように対処しているかをレビューする.

　リスク及び機会の決定に対する組織のアプローチに関する監査は,それ単独の活動として実施しないことが望ましい.それは,マネジメントシステムに関する監査全体において,あら(露)わなものとして行わないことが望ましい.これには,トップマネジメントにインタビューするときを含む.監査員は,次のステップに従って活動し,次の事項のような客観的証拠を集めることが望ましい.

a) inputs used by the organization for determining its risks and opportunities, which may include:
    — analysis of external and internal issues;
    — the strategic direction of the organization;
    — interested parties, related to its discipline-specific management system and their requirements, also;
    — potential sources of risk such as environmental aspects, and safety hazards, etc.
b) method by which risks and opportunities are evaluated, which can differ between disciplines and sectors.

The organization's treatment of its risk and opportunities, including the level of risk it wishes to accept and how it is controlled, will require the application of professional judgement by the auditor.

A.11 Life cycle

Some discipline-specific management systems require the application of a life cycle perspective to their products and services. Auditors should not

a) 組織がそのリスク及び機会を決定するために用いるインプット．これには，次の事項を含めてよい．
　— 外部及び内部の課題の分析
　— 組織の戦略的方向性

　— 組織の分野固有のマネジメントシステムに関係する利害関係者，及びそれらの利害関係者の要求事項
　— 潜在的なリスク源，例えば環境側面及び安全ハザードなど

b) リスク及び機会を評価する方法，これは分野及び業種の間で異なり得る．

　組織のリスク及び機会に関する対応は，組織が受容することを望むリスクのレベル，及びそれをどのように管理するかを含め，監査員による専門的な判断の適用を必要とする．

A.11　ライフサイクル

　分野固有のマネジメントシステムの中には，それらの製品及びサービスに対して，ライフサイクルの視点の適用を必要とするものがある．監査員は，こ

consider this as a requirement to adopt a life cycle approach. A life cycle perspective involves consideration of the control and influence the organization has over the stages of its product and service life cycle. Stages in a life cycle include acquisition of raw materials, design, production, transportation/delivery, use, end of life treatment and final disposal. This approach enables the organization to identify those areas where, in considering its scope, it can minimize its impact on the environment while adding value to the organization. The auditor should use their professional judgement as to how the organization has applied a life cycle perspective in terms of its strategy and the:

a) life of the product or service;

b) organization's influence on the supply chain;

c) length of the supply chain;

d) technological complexity of the product.

If an organization has combined several management systems into a single management system to meet its own needs, the auditor should look carefully at any overlap concerning consideration of the life cycle.

れをライフサイクルアプローチを採用するための要求事項とみなさないことが望ましい．ライフサイクルの視点は，組織がその製品及びサービスのライフサイクルの諸段階にわたってもつ管理及び影響に関する考慮を含む．ライフサイクルの諸段階には，原材料の取得，設計，生産，輸送・配送（提供），使用，使用後の処理及び最終処分を含む．このアプローチによって，組織が，その適用範囲の考慮に当たり，組織に価値を付加しながら，環境への影響を最小化にし得る領域を特定することが可能になる．監査員は，組織がその戦略及び，次の事項に関して，どのようにライフサイクルの視点を適用してきたかについて，専門的な判断を用いることが望ましい．

a) 製品又はサービスの寿命
b) 組織がサプライチェーンに及ぼす影響
c) サプライチェーンの長さ
d) 製品の技術的な複雑さ

　組織が，それ自身のニーズを満たすために，幾つかのマネジメントシステムを一つのマネジメントシステムに組み合わせたならば，監査員は，ライフサイクルの考慮に関する重複がないか，慎重に見ることが望ましい．

A.12 Audit of supply chain

The audit of the supply chain to specific requirements can be required. The supplier audit programme should be developed with applicable audit criteria for the type of suppliers and external providers. The scope of the supply chain audit can differ, e.g. complete management system audit, single process audit, product audit, configuration audit.

A.13 Preparing audit work documents

When preparing audit work documents, the audit team should consider the questions below for each document.

a) Which audit record will be created by using this work document?
b) Which audit activity is linked to this particular work document?
c) Who will be the user of this work document?
d) What information is needed to prepare this work document?

For combined audits, work documents should be developed to avoid duplication of audit activities

A.12　サプライチェーンの監査

サプライチェーンの監査で特定の要求事項に対するものが，必要となり得る．供給者の監査プログラムは，供給者及び外部提供者のタイプに対する適用可能な監査基準に関して策定することが望ましい．サプライチェーンの監査範囲は，例えば，マネジメントシステム全体の監査，単一プロセス監査，製品監査，コンフィギュレーション監査のように異なり得る．

A.13　監査作業文書の作成

監査作業文書を作成するとき，監査チームは各文書に対して次の質問を考慮することが望ましい．

a)　この作業文書を利用してどの監査記録を作成するか．
b)　この特定の作業文書がどの監査活動に結び付くか．
c)　この作業文書を利用するのは誰か．
d)　この作業文書の作成にどのような情報が必要か．

複合監査に対して，監査活動の重複を避けるために，次の事項によって作業文書を策定することが望

by:

— clustering of similar requirements from different criteria;
— coordinating the content of related checklists and questionnaires.

The audit work documents should be adequate to address all those elements of the management system within the audit scope and may be provided in any media.

A.14 Selecting sources of information

The sources of information selected may vary according to the scope and complexity of the audit and may include the following:

a) interviews with employees and other individuals;

b) observations of activities and the surrounding work environment and conditions;

c) documented information, such as policies, objectives, plans, procedures, standards, instructions, licences and permits, specifications, drawings, contracts and orders;

d) records, such as inspection records, minutes

ましい.
— 異なる基準から類似の要求事項をまとめる.

— 関係するチェックリスト及び質問事項の内容を調整する.

　監査作業文書は,監査範囲内のマネジメントシステムの全ての要素に対処するために十分であることが望ましい.また,監査作業文書は,いかなる媒体で提供してもよい.

A.14　情報源の選択
　選択する情報源は,監査の範囲及び複雑さに応じて異なってもよく,それには,次の事項を含んでもよい.

a) 従業員及びその他の人々へのインタビュー

b) 活動並びに周囲の作業環境及び作業条件について行う,観察

c) 文書化した情報.例えば,方針,目的(又は目標),計画,手順,規格,指示,ライセンス及び許認可,仕様書,図面,契約及び注文

d) 記録.例えば,検査記録,会議の議事録,監査

of meetings, audit reports, records of monitoring programme and the results of measurements;

e) data summaries, analyses and performance indicators;

f) information on the auditee's sampling plans and on any procedures for the control of sampling and measurement processes;

g) reports from other sources, e.g. customer feedback, external surveys and measurements, other relevant information from external parties and external provider ratings;

h) databases and websites;

i) simulation and modelling.

A.15 Visiting the auditee's location

To minimize interference between audit activities and the auditee's work processes and to ensure the health and safety of the audit team during a visit, the following should be considered:

a) Planning the visit:
- ensure permission and access to those parts of the auditee's location, to be visited in accordance with the audit scope;

報告書,監視プログラムの記録及び測定結果

e) データの要約,分析及びパフォーマンス指標

f) 被監査者のサンプリング計画に関する情報,並びにサンプリングプロセス及び測定プロセスを管理するためのあらゆる手順に関する情報

g) その他の出所からの報告書.例えば,顧客からのフィードバック,外部の調査及び測定,外部関係者からのその他の関連情報並びに外部提供者の格付け

h) データベース及びウェブサイト

i) シミュレーション及びモデリング

A.15 被監査者の場所の訪問

監査活動と被監査者の作業プロセスとの間の干渉を最小限にするために,また,訪問中の監査チームの安全衛生を確実にするために,次の事項を考慮することが望ましい.

a) 訪問の計画策定
 — 監査範囲に従って訪れる,被監査者の場所の区域への許可及びアクセスを確実にする.

- provide adequate information to auditors on security, health (e.g. quarantine), occupational health and safety matters and cultural norms and working hours for the visit including requested and recommended vaccination and clearances, if applicable;
- confirm with the auditee that any required personal protective equipment (PPE) will be available for the audit team, if applicable;
- confirm the arrangements with the auditee regarding the use of mobile devices and cameras including recording information such as photographs of locations and equipment, screen shot copies or photocopies of documents, videos of activities and interviews, taking into consideration security and confidentiality matters;
- except for unscheduled, ad hoc audits, ensure that personnel being visited will be informed about the audit objectives and scope.

b) On-site activities:

— 監査員に，訪問に差し支えないよう十分な情報を提供する．該当する場合，セキュリティ，健康（例えば，検疫），労働安全衛生上の事項，文化的規範，及び業務時間についての情報である．これには，要請及び推奨されたワクチン接種及び洗浄を含む．

— 該当する場合，必要な個人用保護具（PPE）を監査チームが利用可能であることを被監査者に確認する．

— 情報の記録を含むモバイル機器及びカメラの使用について，セキュリティ及び機密保持の事項を考慮に入れて被監査者との取決めを確認する．情報の記録には，場所及び機器の写真，文書のスクリーンショット又はコピー，活動及びインタビューのビデオなどがある．

— 予定外の臨時の監査を除いて，訪問を受ける要員が，監査の目的及び範囲に関して知らされていることを確実にする．

b) 現地監査活動

- avoid any unnecessary disturbance of the operational processes;
- ensure that the audit team is using PPE properly (if applicable);
- ensure emergency procedures are communicated (e.g. emergency exits, assembly points);
- schedule communication to minimize disruption;
- adapt the size of the audit team and the number of guides and observers in accordance with the audit scope, in order to avoid interference with the operational processes as far as practicable;
- do not touch or manipulate any equipment, unless explicitly permitted, even when competent or licensed;

- if an incident occurs during the on-site visit, the audit team leader should review the situation with the auditee and, if necessary, with the audit client and reach agreement on whether the audit should be interrupted, rescheduled or continued;

附属書 A(参考)

— 運用プロセスに不要な外乱を与えることを避ける.
— (該当する場合)監査チームが PPE を適切に使用していることを確実にする.
— 緊急時の手順(例えば,非常口,集合場所)が伝達されていることを確実にする.

— 中断を最小限にするためにコミュニケーションを行うことを予定する.
— 監査範囲に合わせて,監査チームの規模並びに案内役及びオブザーバの人数を適応させる.これは,実行可能な限り,被監査者の運用プロセスとの干渉を避けるためである.

— 明確に許可されない限り,いかなる機器にも触れたり操作したりしない.これは,たとえその力量があるか,又は資格を保有している場合でもそうである.
— もし現地訪問においてインシデントが生じたならば,監査チームリーダーは,被監査者及び必要な場合は監査依頼者と状況をレビューし,監査の中断,再スケジュール,又は継続のいずれが望ましいかについて合意に至ることが望ましい.

- if taking copies of documents in any media, ask for permission in advance and consider confidentiality and security matters;
- when taking notes, avoid collecting personal information unless required by the audit objectives or audit criteria.

c) Virtual audit activities:
- ensure that the audit team is using agreed remote access protocols including requested devices, software, etc.;

- if taking screen shot copies of document of any kind, ask for permission in advance and consider confidentiality and security matters and avoid recording individuals without their permission;
- if an incident occurs during the remote access, the audit team leader should review the situation with the auditee and, if necessary, with the audit client and reach agreement on whether the audit should be interrupted, rescheduled or continued;
- use floor plans/diagrams of the remote lo-

— もし文書のコピーをとるならば，いかなる媒体であっても，事前に許可を求め，機密保持及びセキュリティ上の事項を考慮する．

— メモをとる場合は，監査目的又は監査基準に求められていない限り，個人情報を収集することを避ける．

c) 仮想監査活動
— 監査チームが，合意した遠隔アクセスプロトコルを用いていることを確実にする．このプロトコルには，要求される機器，ソフトウェアなどを含む．

— もし文書のスクリーンショットを撮るならば，いかなる種類の文書でも，事前に許可を求め，機密保持及びセキュリティ上の事項を考慮し，許可なく個々人を記録することを避ける．

— もし遠隔アクセスにおいてインシデントが生じたならば，監査チームリーダーは，被監査者及び必要な場合は監査依頼者と状況をレビューし，監査の中断，再スケジュール，又は継続のいずれが望ましいかについて合意に至ることが望ましい．

— 遠隔場所のフロアの間取り図・図面を参照と

cation for reference;
- maintain respect for privacy during audit breaks.

Consideration needs to be given to disposition of information and audit evidence, irrespective of the type of media, at a later date, once the need for its retention has lapsed.

A.16 Auditing virtual activities and locations

Virtual audits are conducted when an organization performs work or provides a service using an on-line environment allowing persons irrespective of physical locations to execute processes (e.g. company intranet, a "computing cloud"). Auditing of a virtual location is sometimes referred to as virtual auditing. Remote audits refer to using technology to gather information, interview an auditee, etc. when "face-to-face" methods are not possible or desired.

A virtual audit follows the standard audit process while using technology to verify objective evidence.

して用いる.
— プライバシーに対する尊重を,監査休憩において維持する.

情報及び監査証拠の処分に考慮を払う必要がある.これは,媒体のタイプにかかわらず,後になって,その保持の必要性が無くなったときである.

A.16 仮想活動及び場所の,監査

人々が物理的場所にかかわらずプロセスを実行することを許すオンライン環境(例えば,企業のイントラネット,"コンピューティングクラウド")を利用し,組織が作業を行う場合又はサービスを提供する場合には仮想監査を行う.仮想場所の監査は,仮想監査と呼ばれることがある.遠隔監査は,"対面"方法が可能でないか又は望まれない場合に,情報収集,被監査者へのインタビューなどのための技術を用いることを示す.

仮想監査は,客観的証拠を検証する技術を用いるが,標準的な監査プロセスに従って実施する.被監

The auditee and audit team should ensure appropriate technology requirements for virtual audits which can include:

— ensuring the audit team is using agreed remote access protocols, including requested devices, software, etc.;
— conducting technical checks ahead of the audit to resolve technical issues;
— ensuring contingency plans are available and communicated (e.g. interruption of access, use of alternative technology), including provision for extra audit time if necessary.

Auditor competence should include:

— technical skills to use the appropriate electronic equipment and other technology while auditing;
— experience in facilitating meetings virtually to conduct the audit remotely.

When conducting the opening meeting or auditing virtually, the auditor should consider and the following items:

査者及び監査チームは，仮想監査のための適切な技術要求事項を，確実にすることが望ましい．これには，次の事項を含み得る．
— 監査チームは，必要な装置，ソフトウェアなどを含め，合意された遠隔アクセスプロトコルを使用していることを確実にする．
— 監査に先立って，技術的課題を解決するよう技術的チェックを行う．
— 必要な場合，追加の監査時間をとることを含め，不測の事態への対応計画が，利用可能で伝達されていることを確実にする（例えば，アクセスの妨害，代替技術の利用）．

　監査員の力量には，次の事項を含めることが望ましい．
— 監査を通じて，適切な電子機器及びその他の技術を使用する技術的な技能

— 監査を遠隔で行うために仮想的に会議を進める経験

　初回会議又は監査を仮想的に行う場合，監査員は，次の事項を考慮することが望ましい．

- risks associated with virtual or remote audits;
- using floor plans/diagrams of remote locations for reference or mapping of electronic information;
- facilitating for the prevention of background noise disruptions and interruptions;
- asking for permission in advance to take screen shot copies of documents or any kind of recordings, and considering confidentiality and security matters;
- ensuring confidentiality and privacy during audit breaks e.g. by muting microphones, pausing cameras.

A.17 Conducting interviews

Interviews are an important means of collecting information and should be carried out in a manner adapted to the situation and the individual interviewed, either face to face or via other means of communication. However, the auditor should consider the following:

a) interviews should be held with individuals from appropriate levels and functions performing activities or tasks within the audit

― 仮想監査又は遠隔監査に付随するリスク
― 電子情報の参照又はマッピングのために,遠隔場所のフロアの間取り図・図面を利用すること.
― バックグラウンドノイズによる遮断及び妨害の防止を容易にすること.
― 文書のスクリーンショット又はあらゆる種類の記録をとるために,事前に許可を求め,また,機密保持及びセキュリティ上の事項を考慮すること.
― 監査休憩において,例えば,マイクを消音し,カメラを一時停止するなどし,機密保持及びプライバシーを確実にすること.

A.17 インタビューの実施

インタビューは情報を収集するための重要な手段であり,対面又はその他のコミュニケーション手段を通じて,その場の状況及びインタビュー対象者に合わせた形で行うことが望ましい.ただし,監査員は,次の事項を考慮することが望ましい.

a) インタビューは,監査範囲内で,活動又は業務を遂行している適切な階層及び機能の人々に対して行うことが望ましい.

scope;

b) interviews should normally be conducted during normal working hours and, where practical, at the normal workplace of the individual being interviewed;

c) attempts should be made to put the individual being interviewed at ease prior to and during the interview;

d) the reason for the interview and any note taking should be explained;

e) interviews may be initiated by asking individuals to describe their work;

f) the type of question used should be carefully selected (e.g. open, closed, leading questions, appreciative inquiry);

g) awareness of limited non-verbal communication in virtual settings; instead focus should be on the type of questions to use in finding objective evidence;

h) the results from the interview should be summarized and reviewed with the interviewed individual;

i) the interviewed individuals should be thanked for their participation and cooperation.

b) 通常,インタビューは,通常の就業時間中に,実行可能であれば,インタビュー対象者の通常の就業場所で行うことが望ましい.

c) インタビューを始める前及びインタビュー中に,インタビュー対象者の緊張を解くことを試みることが望ましい.

d) インタビューを行う理由,及びメモをとるのであればその理由を説明することが望ましい.

e) インタビュー対象者の業務について説明を求めることによってインタビューを始めてよい.

f) 質問の種類を注意して選択することが望ましい〔例えば,自由質問,選択質問,誘導質問,価値を認める質問(appreciative inquiry)〕.

g) 仮想設定における限られた非言語コミュニケーションを認識する.その代わりに,客観的証拠を見出すために用いる質問の種類に焦点を当てることが望ましい.

h) インタビューの結果をまとめて,その内容をインタビュー対象者とレビューすることが望ましい.

i) インタビューへの参加及び協力に対して,インタビュー対象者に謝意を表することが望ましい.

A.18 Audit findings

A.18.1 Determining audit findings

When determining audit findings, the following should be considered:

a) follow-up of previous audit records and conclusions;

b) requirements of the audit client;

c) accuracy, sufficiency and appropriateness of objective evidence to support audit findings;

d) extent to which planned audit activities are realized and planned results achieved;

e) findings exceeding normal practice, or opportunities for improvement;

f) sample size;

g) categorization (if any) of the audit findings.

A.18.2 Recording conformities

For records of conformity, the following should be considered:

a) description of or reference to audit criteria against which conformity is shown;

b) audit evidence to support conformity and effectiveness, if applicable;

c) declaration of conformity, if applicable.

A.18 監査所見
A.18.1 監査所見の決定
監査所見を決定するとき,次の事項を考慮することが望ましい.

a) 前回までの監査記録及び監査結論のフォローアップ
b) 監査依頼者の要求事項
c) 監査所見を支持する客観的証拠の正確さ,十分さ及び適切さ
d) 計画した監査活動を実現した程度及び計画した結果を達成した程度
e) 通常の慣行を超える所見,又は改善の機会

f) サンプルサイズ
g) 監査所見の分類(存在する場合)

A.18.2 適合の記録
適合の記録には,次の事項を考慮することが望ましい.

a) 適合を示す監査基準の記述又は監査基準への参照
b) 該当する場合,適合を支持する監査証拠及び,その有効性
c) 該当する場合,適合の明示

A.18.3 Recording nonconformities

For records of nonconformity, the following should be considered:

a) description of or reference to audit criteria;

b) audit evidence;

c) declaration of nonconformity;

d) related audit findings, if applicable.

A.18.4 Dealing with findings related to multiple criteria

During an audit, it is possible to identify findings related to multiple criteria. Where an auditor identifies a finding linked to one criterion on a combined audit, the auditor should consider the possible impact on the corresponding or similar criteria of the other management systems.

Depending on the arrangements with the audit client, the auditor may raise either:

a) separate findings for each criterion; or

b) a single finding, combining the references to multiple criteria.

Depending on the arrangements with the audit cli-

A.18.3　不適合の記録

不適合の記録には，次の事項を考慮することが望ましい．

a)　監査基準の記述又は監査基準への参照
b)　監査証拠
c)　不適合の明示
d)　該当する場合，関係する監査所見

A.18.4　複数の基準に関係する所見への対応

監査中，複数の基準に関係する所見を特定することが可能である．監査員が複合監査の一つの基準に結び付けられた所見を特定する場合，その監査員は他のマネジメントシステムの対応する基準又は類似の基準に対して及ぼし得る影響を考慮することが望ましい．

監査依頼者との取決めに基づき，監査員は次のいずれを提起してもよい．

a)　各基準に対する別々の所見
b)　複数の基準への参照を組み合わせた，一つの所見

監査依頼者との取決めに基づき，監査員はこれら

ent, the auditor may guide the auditee on how to respond to those findings.

の所見にどのように対応するかについて被監査者を導いてよい.

Bibliography

[1] ISO 9000:2015, *Quality management systems — Fundamentals and vocabulary*

[2] ISO 9001, *Quality management systems — Requirements*[1]

[3] ISO Guide 73:2009, *Risk management — Vocabulary*

[4] ISO/IEC 17021-1, *Conformity assessment — Requirements for bodies providing audit and certification of management systems — Part 1: Requirements*

---

[1] See www.iso.org/tc176/ISO9001AuditingPractices Group.

参考文献

[1] **JIS Q 9000**:2015　品質マネジメントシステム—基本及び用語
　　注記　対応国際規格：**ISO 9000**:2015, Quality management systems—Fundamentals and vocabulary（IDT）
[2] **JIS Q 9001**　品質マネジメントシステム—要求事項
　　注記　対応国際規格：**ISO 9001**, Quality management systems—Requirements（IDT）
[3] **JIS Q 0073**:2010　リスクマネジメント—用語
　　注記　対応国際規格：**ISO Guide73**:2009, Risk management—Vocabulary（IDT）
[4] **JIS Q 17021-1**　適合性評価—マネジメントシステムの審査及び認証を行う機関に対する要求事項—第1部：要求事項
　　注記　対応国際規格：**ISO/IEC 17021-1**, Conformity assessment—Requirements for bodies providing audit and certification of management systems—Part 1: Requirements（IDT）

対訳 ISO 19011:2018（JIS Q 19011:2019）
マネジメントシステム監査のための指針［ポケット版］

定価：本体 6,800 円（税別）

2019 年 8 月 26 日　　第 1 版第 1 刷発行

編　　者　一般財団法人 **日本規格協会**

発 行 者　揖斐　敏夫

発 行 所　一般財団法人 **日本規格協会**

　　　　　〒 108-0073　東京都港区三田 3 丁目 13-12 三田 MT ビル
　　　　　https://www.jsa.or.jp/
　　　　　振替　00160-2-195146

製　　作　日本規格協会ソリューションズ株式会社

印 刷 所　株式会社ディグ

© Japanese Standards Association, et al., 2019　　Printed in Japan
ISBN978-4-542-40284-3

- 当会発行図書，海外規格のお求めは，下記をご利用ください．
  JSA Webdesk（オンライン注文）：https://webdesk.jsa.or.jp/
  通信販売：電話（03）4231-8550　FAX（03）4231-8665
  書店販売：電話（03）4231-8553　FAX（03）4231-8667

図書のご案内

ISO 19011:2018
(JIS Q 19011:2019)
マネジメントシステム監査
解説と活用方法

福丸典芳　著

A5判・264ページ　　定価：本体 3,900 円（税別）

【主要目次】
第1章　監査活動の基本
1.1　監査の本質
1.1.1　監査の目的
1.1.2　監査のタイプ
1.2　監査プロセスの問題点
1.2.1　内部監査の構築上の問題点
1.2.2　内部監査の運営上の問題点
1.2.3　第二者監査の運営上の問題点
1.3　監査の仕組み
1.3.1　監査の構造
1.3.2　内部監査に関する要求事項
1.3.3　適合性監査と有効性監査
1.4　監査にかかわる人々の役割
1.5　内部監査，第二者監査，及び第三者審査の関係
1.6　監査員の力量
第2章　ISO 19011 の解説
2.1　"序文"の解説
2.2　箇条1 "適用範囲"の解説
2.3　箇条3 "用語及び定義"の解説
2.4　箇条4 "監査の原則"の解説
2.5　箇条5 "監査プログラムのマネジメント"の解説
2.6　箇条6 "監査の実施"の解説
2.7　箇条7 "監査員の力量及び評価"の解説
第3章　効果的な監査プロセスの構築方法と事例
3.1　監査プロセスの構築方法
3.1.1　業務機能展開の基本
3.1.2　業務機能展開の手順
3.2　内部監査プロセスの業務機能展開の事例
第4章　監査の視点
4.1　単一 MS の内部監査の視点
4.1.1　ISO 9001 の内部監査の視点
4.1.2　ISO 14001 の内部監査の視点
4.1.3　ISO/IEC 27001 の内部監査の視点
4.1.4　ISO 45001 の内部監査の視点
4.2　統合 MS の内部監査の視点
4.3　第二者監査の視点
第5章　監査プログラムの成熟度レベル評価
第6章　監査に関する Q&A

日本規格協会　　https://webdesk.jsa.or.jp/